Angling in India

AF875271

Syed Talia Mushtaq • Ankur Jamwal •
Syed Aalia Mushtaq • Tasaduq Hussain Shah •
Farooz Ahmad Bhat

Angling in India

Exploring Sport Fisheries

Springer

Syed Talia Mushtaq
Division of Fisheries Resource Management
Sher-e-Kashmir University of Agricultural
Sciences and Technology of Kashmir
Ganderbal, Jammu and Kashmir, India

Syed Aalia Mushtaq
Government Degree College
Srinagar, Jammu and Kashmir, India

Farooz Ahmad Bhat
Division of Fisheries Resource Management
Sher-e-Kashmir University of Agricultural
Sciences and Technology of Kashmir
Ganderbal, Jammu and Kashmir, India

Ankur Jamwal
Centre for Climate Change and
Sustainability
Azim Premji University
Bangalore, Karnataka, India

Tasaduq Hussain Shah
Division of Fisheries Resource Management
Sher-e-Kashmir University of Agricultural
Sciences and Technology of Kashmir
Ganderbal, Jammu and Kashmir, India

ISBN 978-981-97-8878-1 ISBN 978-981-97-8876-7 (eBook)
https://doi.org/10.1007/978-981-97-8876-7

© The Editor(s) (if applicable) and The Author(s), under exclusive license to Springer Nature Singapore Pte Ltd. 2024
This work is subject to copyright. All rights are solely and exclusively licensed by the Publisher, whether the whole or part of the material is concerned, specifically the rights of translation, reprinting, reuse of illustrations, recitation, broadcasting, reproduction on microfilms or in any other physical way, and transmission or information storage and retrieval, electronic adaptation, computer software, or by similar or dissimilar methodology now known or hereafter developed.
The use of general descriptive names, registered names, trademarks, service marks, etc. in this publication does not imply, even in the absence of a specific statement, that such names are exempt from the relevant protective laws and regulations and therefore free for general use.
The publisher, the authors and the editors are safe to assume that the advice and information in this book are believed to be true and accurate at the date of publication. Neither the publisher nor the authors or the editors give a warranty, expressed or implied, with respect to the material contained herein or for any errors or omissions that may have been made. The publisher remains neutral with regard to jurisdictional claims in published maps and institutional affiliations.

This Springer imprint is published by the registered company Springer Nature Singapore Pte Ltd.
The registered company address is: 152 Beach Road, #21-01/04 Gateway East, Singapore 189721, Singapore

If disposing of this product, please recycle the paper.

Dedication

To dear students, who daily inspire me with their curiosity and love of the aquatic environment. My dedication to teaching and conservation is fueled by your curiosity about fish biology and your desire to learn more about it.

To all fishing enthusiasts: Thank you for reminding us of the delicate balance that needs to be maintained in our ecosystems through your love of the sport and care for environment.

Let's work together to protect the future of our aquatic habitats and promote the protection of our endemic fish species.

Blest silent groves, oh may you be
For ever mirths best nursery
May pure contents
Forever pitch their tent
Upon these downs, these meads, these rocks, these mountains,
And peace still slumber by these purling fountains
Which we may every year
Meet when we come a fishing here

From The Compleat Angler
Iz Wz

Foreword 1

I am excited to introduce *Angling in India: Exploring the Sport Fisheries*, a significant addition to the sport fishing world. This book showcases the rich aquatic heritage of India, put together with great care and expertise by the authors.

Sport fishing blends fun and nature and plays an important role in India's culture and environment. Our diverse waters, from calm rivers to lively coastal areas, are home to many fish species. Documenting and celebrating this variety is important for both fishing enthusiasts and for preserving our natural heritage. The authors have skillfully balanced sport and conservation, providing insights into the many fish species in India's waters. This book also invites readers to explore the beautiful landscapes and rich ecosystems of India's waters.

I commend the authors for their hard work and dedication. This book is a great resource for anyone who loves angling and the beauty of India's aquatic environments.

ICAR—Central Institute of Fisheries Education
Mumbai, India

C. N. Ravishankar

Foreword 2

The art and science of angling have long been cherished pursuits, intertwining relaxation with the thrill of the catch. The book *Angling in India: Exploring the Sport Fisheries* addresses a significant gap in the literature by offering a comprehensive overview of India and its important sport fish species, the intricate details of fishing equipment, techniques, and the optimal conditions for a successful fishing conditions.

In an era where our aquatic environments are rapidly changing, and tourism is becoming an ever more vital part of our economy, this book underscores the importance of resource management. It presents sport fishing not just as a hobby but as a potential avenue for sustainable tourism development, capable of drawing both domestic and international enthusiasts. The author has meticulously compiled a list of notable fishing locations across India, providing valuable insights into where and how to pursue this rewarding activity. Additionally, by offering sources for fishing gear and information, this book serves also as a practical resource for both novice and experienced anglers.

Sport fishing is more than just a pastime; it is a recreational activity that bridges generations, appealing to a wide range of interests. With the increasing urban population seeking escapes into nature, sport fishing emerges as a family-friendly pursuit that promises relaxation and excitement alike.

This book is a treasure trove of knowledge, packed with essential information and practical advice to guide newcomers through the initially complex world of fishing. The authors adeptly present various fishing methods tailored to different water types and target species, ensuring that readers are well-equipped to embark on their angling adventures. A helpful glossary is included to familiarize readers with essential fishing terminology, making this book an indispensable guide for anyone looking to dive into the world of angling. Whether you are a student, an angler, or someone seeking a new outdoor activity, *Angling in India: Exploring the Sport Fisheries* is your gateway to understanding and appreciating the rich and diverse opportunities for sport fishing in India. May this book inspire you to explore, conserve, and enjoy the abundant aquatic resources of our nation. Indeed this book is a valuable gift for fisheries professionals and anglers as well.

ICAR—Fisheries and Aquaculture Sector Development Planning Dilip Kumar
Mumbai, India

Preface

Fishing, one of humanity's oldest practices, has evolved into a revered sport and a profound connection with nature. *Angling in India: Exploring the Sport Fisheries* delves into the rich and diverse world of sport fishing in India, a country where ancient traditions and vibrant ecosystems intertwine to create a unique angling experience. India, with its vast and varied landscapes, is home to a plethora of freshwater and saltwater fish species. From the serene rivers of the north to the expansive coastal waters of the south, the country's aquatic environment offers unparalleled opportunities for both seasoned anglers and novices alike. This book aims to be a comprehensive guide, capturing the essence of angling in India by highlighting its historical significance, ecological impact, and the sheer joy it brings to those who partake in its timeless activity.

This book primarily targets individuals who are passionate about angling, whether they are experienced anglers or newcomers. It caters to their need for comprehensive information on fishing in India, including details about fish species, techniques, locations, and gear. It also serves as a valuable resource for those concerned about the preservation of aquatic ecosystems in India. It emphasizes responsible angling and the role anglers can play in conservation efforts, making it relevant to environmentalists and conservation advocates. Travelers interested in exploring India's natural beauty and engaging in recreational activities like angling can benefit from this book's insights into fishing destinations, local regulations, and the cultural aspects of angling in India. Scholars and researchers in the fields of ecology, environmental science, and cultural studies may find this book's content relevant for academic purposes and for understanding the cultural and ecological dimensions of angling in India. This book can also be a useful reference for local communities dependent on fishing and for fisheries management authorities interested in promoting sustainable fishing practices and responsible tourism. While the primary audience comprises angling enthusiasts seeking practical guidance, this book's broader appeal extends to those interested in conservation, tourism, academia, and local management of fisheries resources.

The focus of this book, *Angling in India: Exploring the Sport Fisheries*, is on the multifaceted aspects of sport fishing in India. It addresses the rich biodiversity of fish species, the historical and cultural significance of angling, the best fishing sites, and the technical skills required for successful fishing. Moreover, it emphasizes the importance of sustainable practices, conservation efforts, and the role of sport fisheries in promoting ecotourism and supporting local communities.

Ganderbal, India — Syed Talia Mushtaq
Bengaluru, India — Ankur Jamwal
Srinagar, India — Syed Aalia Mushtaq
Ganderbal, India — Tasaduq Hussain Shah
Ganderbal, India — Farooz Ahmad Bhat

Acknowledgments

We acknowledge that the ability to complete this book is a gift from Almighty, and we begin by expressing our gratitude to Him. We extend our profound gratitude and respect to all the authors who contributed chapters to this book. We also thank everyone who assisted with the publication of this book. This endeavor would not have been possible without their continued support and encouragement. Our sincere gratitude goes out to all the authors.

We are deeply thankful to Hon'ble Vice Chancellor SKUAST-K, Prof. Nazir Ahmad Ganaie for his compliments, praise, and blessings to the entire team of young researchers who contributed to this book.

We appreciate the unwavering support and encouragement from the Dean, Faculty of Fisheries, Rangil, Ganderbal at Sher-e-Kashmir University of Agricultural Sciences and Technology in Kashmir.

We are also grateful for the support from our students Dr. Ishrat Mohd, Ifrah Rashid, Khushboo Yaseen, Aziz ul Rehman, Dr. Rinkesh Wanjari, Shahida Mohu-i-din, Misba Rehman, Tanab, Azra Shah, Inab Majeed, and Mir Iqra Farooq and supporting staff, Ali Mohd, Firdous Magray, and Javed Ahmad.

We would also like to extend our heartfelt thanks to Dr. Musadiq Allaqaband and Ubaid Wani who shared their experiences of angling with us and contributed to this book in the form of pictures.

Additionally, we express our sincere gratitude to Springer Publishing Company for publishing this work. Thank you very much.

Finally, we bow down before the Almighty, who has granted us the courage, passion, and strength to complete this effort.

About the Book

This book, *Angling in India: Exploring the Sport Fisheries*, addresses a significant gap by providing a comprehensive overview of the key sport fish species found in Indian waters, along with details about fishing equipment, techniques, and the optimal conditions for fishing. It also discusses resource management in the context of a rapidly changing aquatic environment and a potential for tourism development through sport fishing.

To assist students and anglers, the author has compiled a list of notable fishing locations across the country, as well as sources for fishing gear and angling information. Sport fishing caters to a wide range of interests and serves as a recreational activity for millions of domestic and international tourists. With the increasing urban population, there is a growing demand for outdoor recreational activities, making sport fishing worthy of recognition as a family-friendly pursuit.

Angling is a relaxing and healthy pastime suitable for all ages. This book serves as an essential guide to help newcomers acquire foundational knowledge and embark on an engaging hobby. It is filled with crucial information and practical advice to navigate the initially complex world of fishing. The authors present various fishing methods tailored to different water types and target species. Additionally, a glossary is included to familiarize readers with fishing terminology.

Contents

About the Author

Syed Talia Mushtaq is currently working as an Assistant Professor/Junior Scientist in the Division of Fisheries Resource Management, Faculty of Fisheries, SKUAST-Kashmir, Rangil, Ganderbal (J&K). She graduated in Fisheries Science from the College of Fisheries, G.B.P.U.A.T, Pantnagar, and completed her Master's in Aquaculture from the Central Institute of Fisheries Education, Mumbai. She pursued her PhD in Fisheries Resource Management from the Faculty of Fisheries, SKUAST-K. She has an experience of working as a Farm Manager at the National Fish Seed Farm, Manasbal, for 2 years. She has a number of publications and projects to her credit. She has recently been awarded funding for a project by "The Fisheries Society of the British Isles," UK and for another project by the Wular Conservation and Management Authority, India. She is dedicated to the conservation of endemic fish fauna of the state and the country. Besides, she has also qualified for the NET in the discipline of aquaculture.

Ankur Jamwal is an Assistant Professor at Azim Premji University, Bengaluru, India. Dr. Jamwal has over a decade of experience as an aquatic biologist and ecotoxicologist and has published over 20 research articles in internationally renowned journals. In addition to studying the effects of stress on the physiology of aquatic animals, Dr. Jamwal is also interested in the risk assessment of consumption of food cultivated using water that may be contaminated with toxic elements and compounds. He applies the concepts of aquatic physiology to develop sustainable aquaculture practices. Dr. Jamwal has received multiple awards, including the Best Fisheries Graduate of India (2010) and the ICAR International Fellow. He is a passionate speaker who loves to spread awareness on climate change and the conservation of aquatic ecosystems.

Syed Aalia Mushtaq completed her B.F.Sc., M.F.Sc., and Ph.D. from the Faculty of Fisheries, SKUAST-Kashmir, Rangil, Ganderbal. She has many research publications in national and international journals. She has worked on various aspects of ecology and biodiversity of an important wetland in the state of Jammu and Kashmir (Hokersar), which enjoys the status of being a Ramsar site.

Tasaduq Hussain Shah is currently working as a Professor in the Division of Fisheries Resource Management at the Faculty of Fisheries, SKUAST-Kashmir, Rangil, Ganderbal (J&K) and holds M.F.Sc. and Ph.D. degree from the Central

Institute of Fisheries Education (ICAR-CIFE), Mumbai. He is the recipient of Hiralal Chaudhury Gold Medal for merit during his Master's program and has over two decades of experience in fisheries science. He has guided dozens of master's and doctoral students and published more than 90 research publications in national and international journals. He has also worked in Jammu and Kashmir Fisheries Department in different capacities and has been involved in the propagation and marketing of trout in Kashmir.

Farooz Ahmad Bhat is currently posted as the Dean of the Faculty of Fisheries and Head of the Division of Fisheries Resource Management (FRM), Faculty of Fisheries, SKUAST-K, Rangil, Ganderbal. He has worked on the fish biodiversity, aquatic ecology, population dynamics, distribution patterns of fishes in J&K. He has guided 14 Ph.D., 2 M.Phil, and 41 M.F.Sc. students to date. He has published almost 115 research papers in both international and national journals, with an impact factor as high as 10.92 (Reviews in Aquaculture). He has received several awards at the national and state levels. He has designed various courses in the discipline of fisheries, especially related to coldwater fisheries. For the past nine years, Dr. Bhat has been working as the editor of *SKUAST Journal of Research*.

List of Abbreviations

AIGFA	All India Game Fishing Association
CWS	Coorg Wildlife Society
DCFR	Directorate of Coldwater Fisheries Research
FBE	Fisheries-based ecotourism
ICSF	International Casting Sport Federation
IGFA	International Game Fish Association
IUCN	International Union for Conservation of Nature
MPA	Marine protected areas
MSAA	Maharashtra State Angling Association
PA	Protected areas
SHG	Self-help group
SPF	Sun protection factor
SWOT	Strengths, weaknesses, opportunities, threats
TIES	The International Ecotourism Society
TWFE	Trans World Fishing Expedition
UNESCO	United Nations Educational, Scientific and Cultural Organization
UV	Ultraviolet
UVA	Ultraviolet A radiation
UVB	Ultraviolet B radiation
WASI	Wildlife Association of South India
WHG	Wildlife Help Group
YHG	Youth Help Group

Introduction 1

Angling, or sport fishing, is a timeless pursuit enjoyed by individuals of all ages and backgrounds. Defined as the practice of catching fish using a hook attached to a fishing line and rod, angling has evolved into a sophisticated sport with a rich history and diverse techniques. It offers a unique blend of relaxation, challenge, and connection with nature, making it a beloved activity worldwide, including in India.

The history of angling dates back to early civilizations where fishing was primarily a means of sustenance. References to fishing can be found in the works of Greek philosophers like Aristotle and Plato, as well as in the historical accounts of Plutarch. In India, the tradition of fishing and angling can be traced back to the Ramayana and Mahabharata periods. Inscriptions from the era of King Ashoka indicate that laws existed to protect fish from indiscriminate fishing, underscoring the significance of sustainable fishing practices even in ancient times. During the Mauryan period, Chanakya formulated fishing regulations for emperors, and by the twelfth century, detailed descriptions of fishing techniques, equipment, and the classification of fish were documented in works like King Somesvara's *Manasollasa*. The British colonial era further popularized angling in India, with the introduction of exotic species such as brown and rainbow trout towards the late nineteenth century.

Angling, described by the *Oxford English Dictionary* as "the action or art of fishing with a rod," and by Webster as "the art of fishing with hook and line especially for sport or recreation," has ancient roots. Fishing hooks dating back over 5000 years have been found in the Euphrates basin. Egyptians used barbed hooks around 1200 BC, and earlier wall paintings show scenes of fishing with rods. The Greeks practiced angling, as mentioned by Theocritus in the third century BC, who described bait hanging deceptively from a rod. In fourth century China, metal hooks and silk lines were used, and the oldest known artificial fly, made of wool and feathers, dates back to Macedonia in the third century BC. Angling likely started as a means of obtaining food, similar to hunting, with historical depictions supporting this idea.

© The Author(s), under exclusive license to Springer Nature Singapore Pte Ltd. 2024

S. T. Mushtaq et al., *Angling in India*,
https://doi.org/10.1007/978-981-97-8876-7_1

However, it wasn't until 1496 that angling began to be seen as a sport or recreational activity. This shift is evident in the publication of the *Treatyse of Fysshynge Wyth An Angle*, attributed to Dame Juliana Berners, in the second *Book of St. Albans*. This treatise, aimed at "gentylmen" and "noble men," highlighted the recreational aspect of angling. The sixteenth and seventeenth centuries marked the golden age of recreational fishing, with two significant publications. The first, *The Arte of Angling* by William Samuel, published in 1577, served as a precursor to Izaak Walton's *The Compleat Angler, Or Contemplative Man's Recreation*, published on 9 May 1653. Walton's book established angling as a leisurely pursuit, emphasizing the pleasure derived from the experience itself. Today, recreational fishing has evolved into a highly developed sport enjoyed by many worldwide.

Today's angling methods are diverse, reflecting the sport's evolution through centuries. Common techniques include fly fishing, spinning, and bait fishing, each offering unique challenges and rewards. Anglers use various equipment, including rods, reels, hooks, lines, and baits, which can be either natural or artificial (Fig. 1.1). The classic "hook, line, and sinker" method remains a staple, while modern advancements have introduced sophisticated tools and techniques to enhance the fishing experience. Angling can be pursued from the shore, boats, or launches, employing different methods such as trolling, surf-casting, and still fishing. The sport is also inclusive, welcoming participants regardless of age or gender, and is increasingly recognized for its therapeutic benefits, contributing to its growing popularity.

The right equipment is crucial for successful sport fishing. Anglers must carefully select their gear based on the type of fish they are targeting and the fishing conditions they will encounter. Fishing rods vary in length and action, with lighter rods suitable for trout and heavier rods needed for larger species like marlin. Reels can be either spinning or baitcasting, each offering different advantages depending on the fishing style. Monofilament, fluorocarbon, and braided lines have unique properties, such as strength and visibility, influencing their use in different fishing conditions. Lures and baits, including crankbaits, jigs, and live bait like worms or minnows, are chosen based on the target species and local fishing regulations. Terminal tackle, such as hooks, weights, and swivels, is essential for rigging and presenting bait effectively.

Sport fishing is one of the most adventurous activities worldwide. While many developed and developing countries have advanced tools and techniques for catching fish, India faces a significant gap in information and scientific knowledge about sport fishing, as it is region and species-specific. Although a few states in India adhere to government regulations, there is a lack of awareness about conserving threatened and endangered fish species through sport fishing activities. Additionally, sport fishing can help mitigate unemployment by generating income and promoting ecotourism. Government involvement is essential for the sustainable development and management of sport fisheries in India.

India is home to a variety of game fish, offering anglers the thrill of catching species that put up a spirited fight. The Mahseer, a large carp native to the rivers of the Indian subcontinent, is particularly renowned among anglers. The Golden Mahseer, found in the Himalayan region, is especially prized for its size and challenge. Other

Fig. 1.1 An angler fishing in a mountain stream of Kashmir. Picture credit: Ubaid Wani

notable species include the Goonch catfish, various types of trout (introduced by the British), and numerous saltwater fish such as mackerel, marlin, and tuna.

India boasts numerous prime locations for sport fishing, each offering unique experiences and opportunities. The Lower Himalayas, with rivers like the Ramganga and the Sarda, are celebrated for Mahseer fishing. In Karnataka, the Kaveri River is famed for its abundant Mahseer population. The rivers and streams of the Nilgiris Hills in Tamil Nadu and the streams around Munnar in Kerala are excellent spots for trout fishing. In the northern regions, the Indus, Jhelum, and Lidder rivers in Jammu and Kashmir, along with the Ganges and Yamuna rivers and their tributaries, are rich in Mahseer, trout, and Goonch. The coastal regions of the Andaman and Lakshadweep Islands offer opportunities to catch various saltwater species, making them popular among anglers seeking diverse fishing experiences.

Sport fishing plays a vital role in promoting environmental awareness and conservation. Anglers often develop a deep appreciation for aquatic ecosystems, leading to advocacy for sustainable practices and habitat preservation. Additionally, sport fisheries contribute significantly to local economies through tourism, creating jobs and supporting businesses related to fishing gear, guides, and accommodations. The economic impact of sport fisheries can be substantial. In many regions, recreational fishing attracts tourists who spend money on licenses, equipment, lodging, and food. This influx of visitors can provide essential income for local communities, particularly in rural areas where alternative economic opportunities may be limited.

Sport fisheries also promote environmental stewardship. Many anglers participate in conservation efforts, such as catch-and-release practices, habitat restoration projects, and advocacy for cleaner waterways. By fostering a connection between people and nature, sport fishing can lead to increased public support for environmental protection initiatives.

As angling grows in popularity, there is an increasing emphasis on sustainable practices to protect aquatic ecosystems and fish populations. Over-fishing, marine pollution, and the impacts of fish farming pose significant challenges. To address these issues, regulations such as size limits and catch-and-release practices are implemented to ensure the long-term viability of fish populations. Anglers are encouraged to adhere to these guidelines, balancing their enjoyment of the sport with ecological responsibility.

Angling in India, with its deep historical roots and modern advancements, offers a rich and rewarding experience for enthusiasts. Whether for the thrill of the catch, the serenity of nature, or the joy of mastering diverse techniques, angling continues to captivate and inspire, contributing to both personal fulfillment and the conservation of aquatic life. Sport fisheries offer a unique blend of recreation, economic opportunity, and environmental stewardship. By understanding the significance of sport fishing, the species involved, and the challenges faced, stakeholders can work together to promote sustainable practices that benefit both anglers and aquatic ecosystems. In India, with its rich biodiversity and stunning landscapes, sport fishing holds the potential to become a thriving industry that not only enriches the lives of individuals but also contributes to the conservation of the country's natural heritage. Through responsible management and community engagement, sport fisheries can continue to flourish for generations to come.

Given the growing significance of sport fishing in recent years, this book aims to provide readers with comprehensive information on the current status of sport fisheries in India. It covers the natural history of key freshwater and saltwater game fish, sport fishing equipment, and detailed guidance on where, when, and how to catch game fish. Additionally, it discusses the management of sport fisheries, the role of tourism in promoting sport fishing, and highlights important fishing locations.

History of Sport Fishing in India

2

Abstract

Sport fishing has a long and illustrious history in India and can be traced back to ancient Greek and Egyptian writings, as well as the writings of Aristotle and Plato. The origins of sport fishing as a recreational activity are unclear, but it probably gained popularity in England after the Norman conquest in 1066. In India, the sport began during the British colonial era when anglers from all over the world came to fish for highly sought-after species like the mighty mahseer.

Keywords

Sport fishing · Recreational angling · Mahseer · Fisheries management · Tourism

1 Introduction

Fishing is one of humanity's oldest pursuits, blending skill and scientific understanding of fish behavior. The term "angling" originates from the Greek word "ankos glen" meaning "barbed hook" and the Old English word "anga hook" meaning "angled hook" (Fig. 2.1). Early humans, through observation, gained knowledge about fish and used hooks and lines for sustenance. As time progressed, the advent of nets and metal hooks expanded fishing to all social classes, transforming it into a popular pastime. In India, archaeological evidence from Mohenjo-daro, Harappa, and other Indus Valley sites suggests that angling has been practiced since the third and second millennia B.C., reflecting its deep historical roots.

© The Author(s), under exclusive license to Springer Nature Singapore Pte Ltd. 2024
S. T. Mushtaq et al., *Angling in India*,
https://doi.org/10.1007/978-981-97-8876-7_2

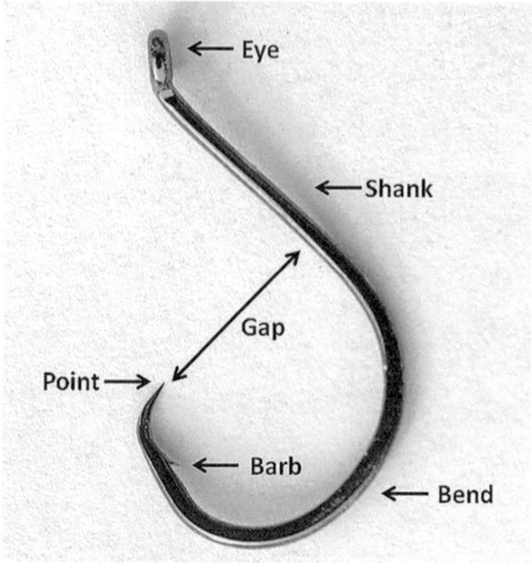

Fig. 2.1 Fish hook. Mike Cline, CC BY-SA 3.0. http://creativecommons.org/licenses/by-sa/3.0/. Via Wikimedia Commons

2 Evidences of Angling During the Indus Valley and Harappan Civilization

Marshall (1931) and Mackay (1930) documented 16 fish-hooks from Mohenjo-daro, all of which were barbed. The Harappa report lists only one unbarbed hook. Chanhu-daro yielded seven specimens, three of which are barbed and four unbarbed, as recorded by Majumdar (1934). Harappa might not have favored angling due to the characteristics of the Ravi and Indus rivers, with the latter likely being a richer fishing resource. However, based on research on the fisheries of the Ravi at Lahore, the river was quite rich in fish. Therefore, the scarcity of fish-hooks at Harappa cannot be attributed to a lack of fish (Fig. 2.2).

Sarkar (1953) examined fish-hooks found at Indus Valley sites, noting their striking similarity to modern hooks. He also compared the use of these fish-hooks among India's primitive tribes with those found in other Middle Eastern and Egyptian excavations. He further classified Indus Valley fish-hooks into two main types: barbed and unbarbed.

Overall, the evolution of angling seems to be more closely linked to inland waters, particularly impounded ones, while tidal waters were fished using nets, traps, and baskets. It was concluded that the Indus Valley developed the best metallic fish-hooks, surpassing those at similar sites in Egypt and Mesopotamia, and showing continuity with modern hooks as noted in pottery designs by Mackay (1930).

Fig. 2.2 Prehistoric hook with sinker. Pedergnana A, Cristiani E, Munro N, Valletta F and Sharon G, CC BY-SA 4.0. https://creativecommons.org/licenses/by-sa/4.0. Via Wikimedia Commons

3 Evidences of Angling in the Historical Period

Vedic literature frequently mentions fishing with nets but lacks references to angling. It wasn't until the early twelfth century, nearly 3000 years post-Harappa, that the earliest known text on angling appeared in the form of *Manasollasa* (1127 AD), written by the Chalukya king Someśvara (Hora, 1951a).

Someśvara's kingdom spanned the Godavari and Krishna watersheds and included both western and eastern Indian coasts. His encyclopedic work *Manasollasa* is divided into five sections called prakaranas. The fourth and fifth prakaranas focus on various amusements and recreations, including a chapter titled Matsyavinoda ("Pastime of Fishing"). This chapter details how a king can enjoy angling, listing 37 species of Indian sporting fish categorized into marine, freshwater, and anadromous types. Someśvara discusses fishing tackle in three main components: line, rod, and hook, and offers specific ground-bait recipes and feeding methods for various fish species. His hints on fishing techniques cover ground-baiting, tackle, floats, baits, casting the line, detecting fish bites, striking, and playing a fish.

By the twelfth century, angling in peninsular India had reached a high standard, with methods aligning closely with those used by modern anglers in India. This sudden and detailed account of angling is further explained by old irrigation tank inscriptions in peninsular India, indicating that pond-culture and angling were already advanced when Someśvara wrote *Manasollasa*. Inscriptions from the fifth and sixth centuries A.D. show that irrigation tanks were maintained from paddy cultivation revenue, evolving to include fishing revenue for tank maintenance by the eleventh century. This historical narrative suggests that the revival of sport fishing in southern India began around the middle of the tenth century.

4 Angling During the British Rule

The British played a significant role in the development of angling in India, introducing new techniques and species. They introduced the concept of recreational fishing, transforming it into a popular leisure activity. The gypsies of Europe, known to use fragments of Mongolian, Hindi, and other Asiatic languages, employed angling methods similar to those described in the Matsyavinoda. This continuity suggests that the gypsies may have wandered through India, contributing to the cultural exchange of angling practices (Hora, 1951b).

5 Inception of Sport Fisheries in Indian Context

The history of angling dates back to ancient times, with early mentions in Greek philosophy, Roman history, and ancient Indian texts. Recreational fishing likely reached England with the Norman conquest in 1066 (McCully, 2000) and became more prominent with the publication of Izaak Walton's *The Compleat Angler* in 1653, highlighting fishing as a leisure activity. In India, King Someśvara's *Manasollasa* (1127 AD) was one of the earliest texts to mention recreational fishing. Mahseer (Tor spp) were first identified in the Ganges in the early nineteenth century and gained popularity through the Oriental Sporting Magazine (Mandal et al., 2018). Notable works by H. S. Thomas and Edward James "Jim" Corbett further popularized angling in India, with books like Dhu's *The Angler in India* on the Mighty Mahseer (1923) and A. St. J. MacDonald's *Circumventing the Mahseer and Other Sporting Fishes* (1948).

The British introduced brown trout in the 1860s and rainbow trout in 1909 to the streams and rivers of the Himalayas and Western Ghats, significantly boosting angling activities (Sehgal, 1999). Post-independence, recreational angling in India saw a decline but regained interest in the 1970s, with both foreign and Indian anglers focusing on Indian freshwater systems. Various angling organizations, including the All India Game Fishing Association, played key roles in promoting the sport, with regional affiliates like Mustad, Maharashtra State Angling Association (MSAA)-Powai, SHIMANO, The Rogue Angler, Anglers Association (Futala Lake)-Nagpur, Gypsy's Camp-Bhakhra, and Dasvino Town & Country Club working towards popularizing angling across India (Mandal et al., 2018). A 24 km stretch of the Ramganga River that is contained under the borders of Uttarakhand's Corbett National Park was leased from the Forest Department in 2004 by a group of local stakeholders from the Mahseer Conservancy. The Conservancy's goals were to support the preservation of the golden mahseer, *Tor putitora*; draw recreational anglers to the area; use the money made from these trips to finance conservation initiatives; and benefit the local communities both socially and economically. Furthermore, through his television shows "Jungle Hooks India" and "River Monsters," world-famous recreational angler Jeremy Wade contributed to the promotion of the mahseer and the goonch catfish (*Bagarius bagarius*) as significant fishing species. The Indian Fisheries Act's special rules allowed brown and rainbow trout to be caught in the Himalayan region of northern India on rod and line using artificial and live baits; nevertheless, the fishing season, bag limit, and approved baits were regulated (Sehgal, 1999).

6 Present Status of Angling

The Supreme Court of India issued an order in July 2012, prohibiting recreational fishing in all of the nation's protected areas. Currently, the bulk of foreign and Indian recreational fishermen focus their efforts on river reaches outside of protected areas (PAs) because this is the only area in India where recreational fishing is allowed. Recreational anglers in India primarily target the following species: *Neolissochilus hexagonolepis* (copper or chocolate mahseer), *Gibelion catla* (Indian major carp/catla), *T. putitora* (golden mahseer), *T. khudree* (Deccan mahseer), and *Tor* sp. (formerly known as *Tor mussullah* or the hump-backed mahseer. The number of Indian anglers is reportedly on the rise, despite the fact that the number of foreign anglers visiting the Indian freshwater systems has significantly decreased since the angling ban. This could be the reason behind increases in expenses related to fishing. For instance, overseas businesses have expressed a strong desire to enter the Indian tackle industry, while Indian tackle companies claim notable sales growth.

State fisheries departments oversee fishing in the northeastern states, while state forest departments oversee fishing in the northwestern states. Nevertheless, there are no laws governing the quantity of rods in the northeastern states. The appropriate State Fisheries Department issues licenses to vessels operating in maritime waters.

Recreational fishing in marine waters has been a very popular pastime in recent years. The Indian Ocean and the Bay of Bengal (particularly the waters surrounding the Andaman Islands) have become popular destinations for domestic anglers due to the efforts of numerous angling associations. Some of the most popular associations of the country include All India Game Fishing Association/AIGFA, International Game Fish Association/IGFA, and All India Indian Angler. Other state based associations are Wildlife Association of South India/WASI and Coorg Wildlife Society/CWS (Karnataka), Maharashtra State Angling Association/MSSA and Anglers Association, Futala Lake (Maharashtra), Chennai Anglers Association and Annamalai Anglers Association (Tamil Nadu), Cochin Anglers (Kerala), Jamshedpur Anglers (Jharkhand), Kolkata Anglers (West Bengal), Sikkim Anglers Association (Sikkim), Assam (Bhoreli) Angling & Conservation Association/A(B)ACA (Assam), Nagaland Anglers Association (Nagaland), The Himalayan Outback (Uttarakhand), Tripura Angling Association (Tripura), Trout Conservation and

Angling Association (Himachal Pradesh), Kemang Angling Association and Pasighat Angling Club (Arunachal Pradesh).

During the fishing season, which normally runs from October to April depending on the arrival of the monsoon, about 90–120 boats (capable of holding five to six people) operate each month in the marine waters. Caranx ignobilis (giant trevally), Gymnosarda unicolor (dogtooth tuna), Lates calcarifer (Asian seabass), Sphyraena sp. (barracuda), Rachycentron canadum (cobia), and Thunnus obesus (bigeye tuna) are among the marine species that Indian anglers pursue.

7 Conclusion

The history of sport fishing in India is a testament to its enduring appeal and cultural significance. From its ancient origins in the Indus Valley to the sophisticated practices of the twelfth century and the innovations introduced during British rule, angling has evolved into a beloved recreational activity. Today, with the efforts of various organizations and enthusiasts, sport fishing continues to thrive, offering unparalleled enjoyment and fostering a deeper appreciation for India's rich aquatic biodiversity setting the stage for further exploration of its modern developments and future prospects.

References

Hora, S. L. (1951a). Maintenance of irrigation-tanks through fishery-revenue in ancient India. *Journal of the Asiatic Society: Letters, XVII*, 41–50.

Hora, S. L. (1951b). Knowledge of the ancient Hindus concerning fish and fisheries of India. *Journal of the Asiatic Society: Letters, 17*, 145–169.

MacKay, E. (1930). Painted pottery in modern Sind: A survival of an ancient industry. *The Journal of the Royal Anthropological Institute, LX*, 127–135.

Majumdar, N. G. (1934). *Explorations in Sind: Being a report of the exploratory survey carried out during the years 1927-1928, 1929-1930, 1930-1931* (p. 48). Memoirs of the Archaeological Survey of India.

Mandal, A., Das, S. K., Biswas, B., & Kairnar, S. O. (2018). Future scope of sport fisheries development in India: a review. *Indian Journal of Animal Health, 57*(1), 1–16.

Marshall, J. (1931). *Mohenjo-daro and indus civilization* (Vol. 3). Arthur Probsthain.

McCully, C. B. (2000). *The language of fly-fishing* (pp. 76–78). Taylor & Francis.

Sarkar, H. (1953). Fish-hooks from the Indus valley. *Journal of Asian Social Science Research, XIX*, 133–139.

Sehgal, K. L. (1999). Coldwater fish and fisheries in the Indian Himalayas: Rivers and streams. In *Fish and fisheries at higher altitudes: Asia* (p. 304). FAO.

Principle Game Fishes of Inland Waters

3

Abstract

Anglers from all over the world come to India's inland waters, which are made up of rivers, lakes, and reservoirs, because they are home to a wide variety of game fish. The most well-known species are the Brown Trout (*Salmo trutta fario*), Rainbow Trout (*Oncorhynchus mykiss*), Deccan Mahseer (*Neolissochilus hexagonolepis*), Golden Mahseer (*Tor putitora*), and Tor Mahseer (*Tor tor*). These fish are highly valued for their size, table quality, and fierce nature. The greatest times to fish differ depending on the location; in the Himalayan rivers, October to November and mid-February to mid-May are the most fruitful times, whereas in South India, April to September is the finest time. Overfishing, pollution, and habitat degradation are some of the threats facing these game fish populations, which emphasizes the necessity of sustainable management techniques to guarantee their long-term existence and the continuation of this popular fishing sport in India.

Keywords

Game fishes · Mahseer · Trout · Angling · Sport fishing · Inland waters · Rivers · Lakes

1 Introduction

India is a well-liked location for fishermen and fishing enthusiasts due to its varied selection of freshwater game species. The Mahseer, Golden Mahseer, and Goonch catfish are just a few examples of the famous species. Because of their strength and exciting fighting prowess, these game fish give anglers an unforgettable experience. Some Indian freshwater game fish are important for culture and ecology in addition to being good athletes. For instance, the Mahseer is frequently referred to as the

© The Author(s), under exclusive license to Springer Nature Singapore Pte Ltd. 2024

S. T. Mushtaq et al., *Angling in India*,
https://doi.org/10.1007/978-981-97-8876-7_3

"tiger of the river" due to its enormous head and stunning size. Anglers from all over the world, both locally and outside, highly prize it. In addition, the Brahmaputra drainage system in northeastern India is regarded as a hotspot for freshwater fish variety and one of the biggest hydrographic basins in Southeast Asia (Baishya et al., 2016).

These freshwater game fish contribute to India's rich aquatic biodiversity in addition to making the country's waterways more enjoyable for recreational purposes. The conservation of these game fish is essential to the general well-being and sustainability of India's freshwater ecosystems, since they contribute significantly to the biological balance of their various habitats. Additionally, the economy benefits from the commercial exploitation of these game fish since fishing tourism draws tourists from both domestic and foreign locations. All things considered, India's freshwater game fish are an invaluable and beloved resource that offer both ecological significance and pleasure for entertainment. In conclusion, India's freshwater game fish contribute to the natural and cultural heritage of the nation. To maintain their numbers and guarantee that people can continue to enjoy this special resource, it is essential that they be properly managed and conserved.

These sport fishes are unique and the distinct physical characteristics and behaviors of each species distinguish them from one another and add to the diversity of freshwater fish populations in India. To ensure that these species continue to exist for the benefit of future generations of ecologists and anglers, successful conservation requires accurate identification and knowledge of these species. The distinguishing characters of some of the well-known cold water fish species have been presented as follows:

1.1 *Tor putitora* (Hamilton, 1822)

Common name: Putitor Mahseer, Himalayan Mahseer, Golden Mahseer (Fig. 3.1).

The Golden Mahseer, or *Tor putitora*, is one of India's most recognizable and sought-after freshwater game fish (Pant et al., 2018). It is renowned for both its enormous size and its strength and fighting ability. It is considered as culturally significant because of its huge head and prestige as the "great tiger" of the river, as indicated by its native name: Mahseer. However, there is concern over the declining numbers of Golden Mahseer and efforts are being made to conserve the population.

Distinguishing characters: It has a streamlined body, oblong and somewhat compressed with broad scales that produce a golden-brass tint (Jhingran, 1959; Talwar & Jhingran, 1991). The distinctive features of this species include a huge, projecting head, an extended snout, and a tiny, crescent-shaped mouth with two pairs of barbels. In addition, the teeth are either nonexistent or very primitive (Harrison, 1963; Menon, 1999). Along with the ventral fins red to orange colour, it is noted that the larger dorsal fin is located halfway along the back, making it a distinguishing characteristic (Pethiyagoda, 1991; Rainboth, 1996). The Mahseer can swim quickly because of its broad, forked tail, especially in turbulent waters (Campana, 1992). The length of this species can exceed 3 m. The lateral line is scarcely distinguishable. It is classified as a benthopelagic feeder because it mostly eats aquatic insects, crustaceans, and tiny fish (Campana, 1992; Menon, 1999).

1.2 *Tor tor* (Hamilton, 1822)

Common name: Tor Mahseer, Deep bodied Mahseer, Red fin Mahseer, Tor Barb (Fig. 3.1).

Distinguishing characters: It has a deep body, with back more arched than the lower profile. Upper lip is capable of being extended beyond the lower one to form

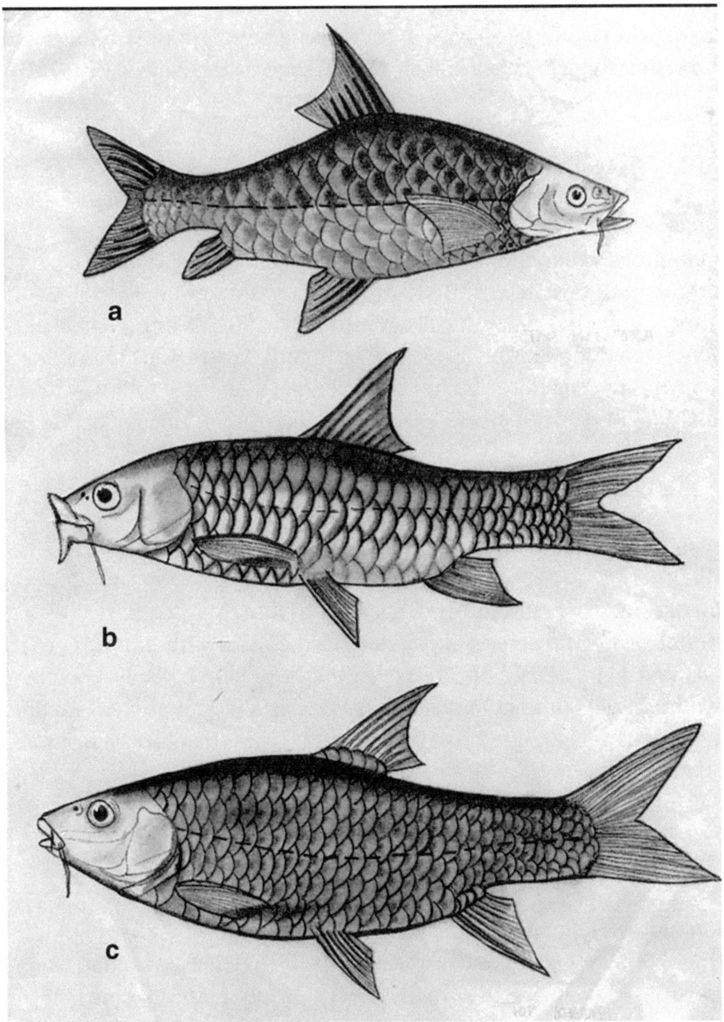

Fig. 3.1 Game fishes: (**a**) *Tor putitora*; (**b**) *Tor tor*; (**c**) *Tor khudree*

a cup. Rostral pair of barbels are slightly smaller than the maxillary pair. Pectoral fins are long and reach up to ventral fins. Young specimen have pink sides and a greyish-greenish top. Adults have a pale olive-green bottom and a greenish-gold top (Sehgal, 1987; Talwar & Jhingran, 1991).

1.3 *Tor khudree* (Sykes, 1839)

Common name: Deccan Mahseer, Khudree Mahseer, Black Mahseer (Fig. 3.1).

Distinguishing characters: The dorsal and ventral body profiles of the fish are moderately convex. Upper jaw is slightly longer than the lower jaw with thick lips. Lower lip is provided with a median lobe. Two pairs of barbels are present, the maxillary pair reaching far behind eyes. It has a silvery-bluish colour, with blood red fins or fins tipped with a bluish tinge (Day, 1889; Sehgal, 1987; Talwar & Jhingran, 1991).

1.4 *Tor chelynoides* (McClelland, 1839)

Common names: Chitrahtu, Dark Mahseer, Naziritor Chelynoides.

Distinguishing characters: Body of the fish is somewhat elongated, giving the appearance of equally convex dorsal and ventral profiles. While snout overhangs the upwardly directed mouth, the upper jaw is slightly larger. Lips are relatively thick without any lobes. The lower lip has a continuous transverse fold without any lobe. Two equal-length pair of barbels are present. The lateral line is complete (Sehgal, 1987; Shahi et al., 2023).

1.5 *Acrossocheilus hexagonolepis* (McClelland, 1839)

Common name: Copper Mahseer, Chocolate Mahseer (Fig. 3.2).

Distinguishing characters: It is a deep bodied fish with thin lips covered with tubercles. Two pairs of barbels are present. Dorsal side of body is olive green in colour with silvery white below and a golden yellow lateral band running in the middle. Fins are slate grey in colour with pale margins (Laishram, 1991; Sehgal, 1987).

1.6 *Wallago attu* (Bloch & Schneider, 1801)

Common name: Helicopter Catfish (Fig. 3.2).

Distinguishing characters: Head is broad with a depressed snout. Body is elongate and strongly compressed. Mouth is very deeply cleft, its corner reaching far behind eyes. Teeth in jaws are set in wide bands with vomerine teeth in two small patches. Two pairs of barbels are present; maxillary barbels extending to anterior margin posterior of anal fin, mandibular barbels extending to angle of mouth. Eyes are small with a free orbital margin. Dorsal fin is small, while the anal fin is very long. Mandibular barbel is longer than the pelvic fin (Kottelat, 2001; Rainboth, 1996; Sehgal, 1987; Taki, 1974).

1.7 *Oncorhynchus mykiss* (Walbaum, 1792)

Common name: Rainbow Trout (Fig. 3.2).

Distinguishing characters: It has a slender and elongated body. Many irregular spots are present on head, body, dorsal fin, and tail. An iridescent pink or red lateral line is present. The adipose fin is present along the back near the tail. Dorsal fin has

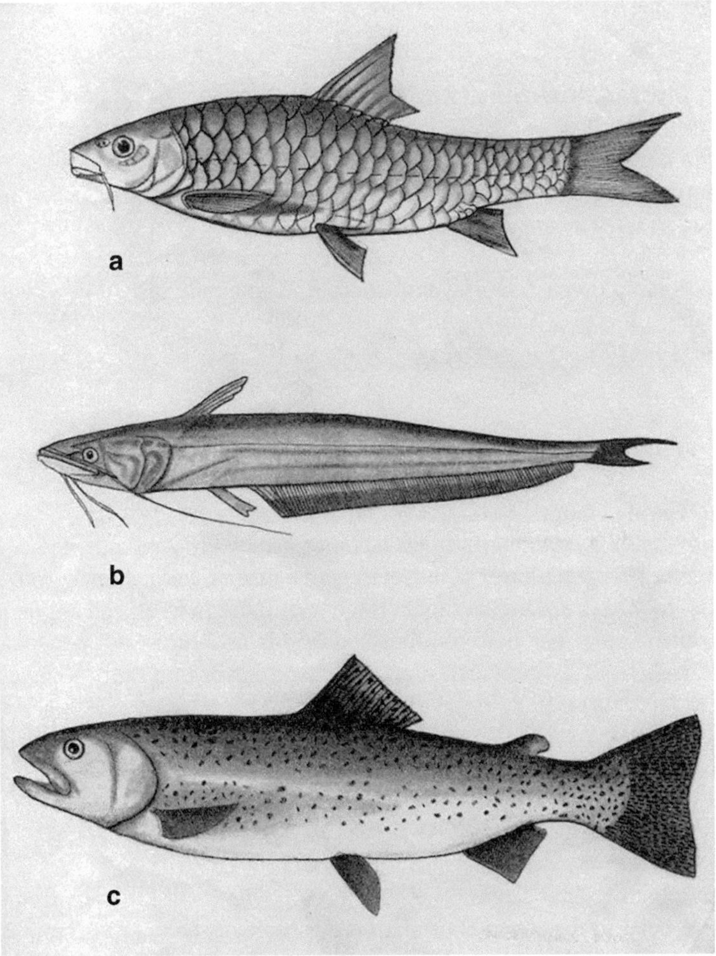

Fig. 3.2 Game fishes: (**a**) *Acrossocheilus hexagonolepis*; (**b**) *Wallago attu*; (**c**) Oncorhynchus mykiss

soft rays and all fins are entirely without spines. Colour is silvery or blue-green with pink belly and clear or red-orange fins (Kottelat & Freyhof, 2007; Spillman, 1961).

1.8 *Raimas bola* (Hamilton, 1822)

Common name: Trout Barb, Indian Trout (Fig. 3.3).

Distinguishing characters: It has an elongated, slender body laterally compressed with a sharp snout. Juvenile fish possess a pair of rudimentary maxillary barbels which are absent in adults. Small scales (85–95) are present along its lateral line. Dorsal part of the body is greenish black, separated from the silvery flanks by a golden stripe running along the length of the body. Yellowish fins with greenish blue spots on body are also seen (Rahman, 1989; Shafi & Quddus, 2001; Talwar & Jhingran, 1991).

1.9 *Mystus/Aorichthys aor* (Sykes, 1822)

Common name: Tengra Catfish (Fig. 3.3).

Distinguishing characters: It has a compressed and elongated body with a large and broad snout. The upper jaw is longer with thin lips, jaws sub equal, teeth villiform, with four pairs of barbels. Median longitudinal furrow on head extends to the base of occipital process. Dorsal surface of head and body is light grey with bluish and golden reflection and a white belly (Day, 1878; Gupta, 2015a; Talwar & Jhingran, 1991).

1.10 *Mystus (Aorichthys) seenghala* (Sykes, 1822)

Common name: Tengra, Catfish (Fig. 3.3).

Distinguishing characters: It has a compressed body with a depressed head, a longer upper jaw, a shallower mouth cleft, and a median longitudinal groove on the head. The rugose is as long as snout. The dorsal spine is weak and strong pectoral with serrated spine. The base of adipose dorsal is as long as or even longer than rayed dorsal. Body is brownish grey above and silvery on sides (Sehgal, 1987; Talwar & Jhingran, 1991).

1.11 *Channa marulius* (Hamilton, 1822)

Common name: Murrels, Snakeheads (Fig. 3.4).

Distinguishing characters: The body is sub-cylindrical. It tapers from the flat snake-like head to the rounded caudal fin. Multiple cephalic pits are present. The maxilla extends beyond the eye. Young ones have a brilliant orange lateral band and adults have 5–6 cloudy bands below lateral line. A large black ocellus is present at

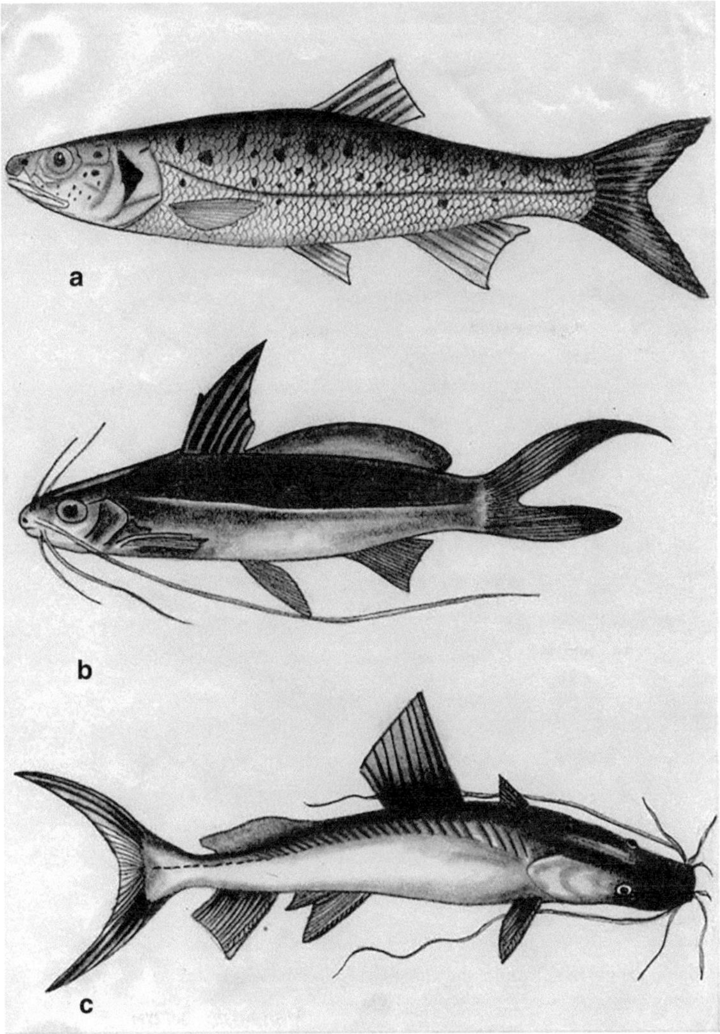

Fig. 3.3 Game fishes: (**a**) *Raimas bola*; (**b**) Mystus aor; (**c**) Mystus seenghala

the upper part of the base of caudal fin (Courtenay & Williams, 2004; Fuller, 2011; Sehgal, 1987).

1.12 *Channa striatus* (Bloch, 1793)

Common name: Murrels, Snakeheads (Fig. 3.4).

Distinguishing characters: The sub-cylindrical body tapers from the flat snake-like head to the rounded caudal fin. Lower jaw is longer and maxilla reaches up to

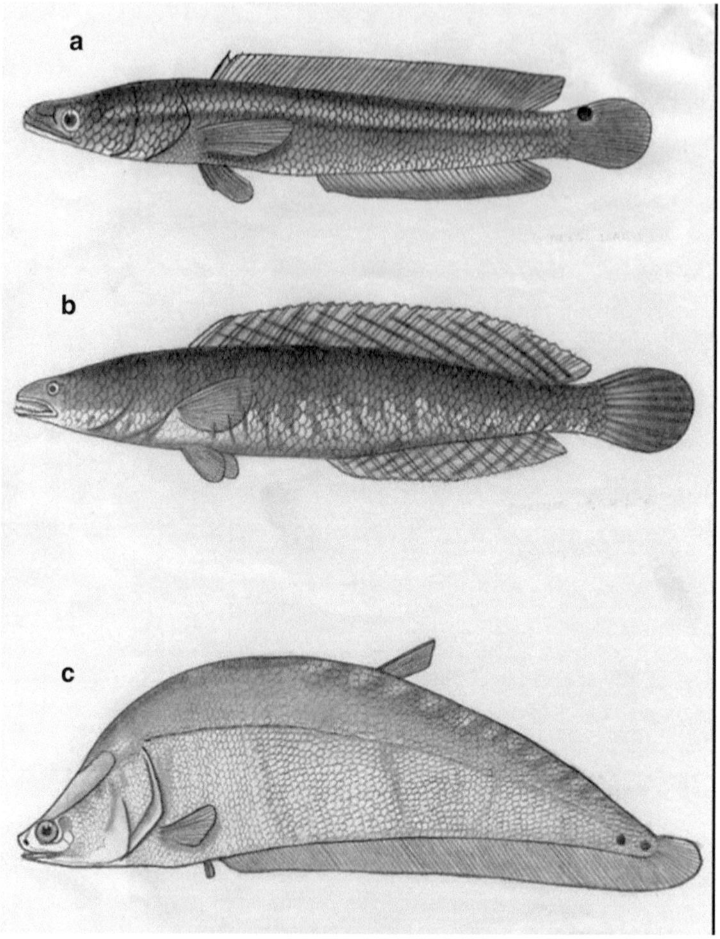

Fig. 3.4 Game fishes: (**a**) *Channa marulius*; (**b**) *Channa striatus*; (**c**) *Notopterus chitala*

below hind border of eye. Yellow or orange abdomen with dark brown dorsal side. A large back ocellus at upper part of the base of caudal (Allen, 1991; Sehgal, 1987; Yamamoto & Tagawa, 2000).

1.13 *Notopterus chitala* (Hamilton, 1822)

Common name: Chital (Fig. 3.4).

Distinguishing characters: The body is compressed body with a small head and a flat abdomen. It has a semi-circular dorsal curve with a notch at the point of insertion of eye. Mouth cleft extends up to the middle of the eye. Dorsal fin is small, situated almost in the central part of the body. Anal fin is long confluencing with the caudal. Body is silvery grey at the back with a number of dark bands (Day, 1878; Rahman, 1989; Sehgal, 1987).

1.14 *Schizothoraichthys esocinus* (Heckel, 1838)

Common name: Chirruh.

Distinguishing characters: It has a sub-cylindrical body with longer upper jaw. Gap of mouth extends up to middle part of eye. Labial fold is interrupted in the middle. Two pairs of barbels are present. Dorsal spine is serrated and is equal to the length of head. Body is silvery with numerous black spots located dorsally (Kullander et al., 1999; Sehgal, 1987).

1.15 *Schizothorax planifrons* (Heckel, 1838)

Common name: Chush.

Distinguishing characters: Body is narrow and elongated. Head is elongated, flat above and arched below. An anterior, oblique, and wide mouth with thick lips is present. The lower lip is broadly interrupted in the middle. Jaws are almost equal in length. Ventral fins are little shorter than pectorals. Caudal fin is deeply furcate with pointed lobes. Body has a darker dorsal surface with ventral surface being powdered with black pigments (Sehgal, 1987).

1.16 *Clupisoma garua* (Hamilton, 1838)

Common name: Garua (Fig. 3.5).

Distinguishing characters: Body is elongated and compressed with keeled abdomen between pelvic and vent. Upper jaw is longer. Width of gap of mouth is 2/5 in head length. Four pairs of barbels are present with maxillary pair extending up to middle of pelvic origin. Dorsal spine is slender and serrated as long as head. Pectoral spine is stronger and serrated as long as dorsal spine. Adipose dorsal is present in young ones but absent in the adult. Body is silvery-grey above, lighter below with fins tinted grey (Sehgal, 1987; Talwar & Jhingran, 1991).

1.17 *Eutroplichthys vacha* (Hamilton, 1822)

Common name: Buchwa (Fig. 3.5).

Distinguishing characters: Snout compressed and pointed. Opening of the mouth oblique extending beyond the mid-portion of the eye. Upper jaw slightly longer. Barbels four pairs, maxillary pair reaching almost to the operculum; nasal pair reaching beyond the length of head. Dorsal spine thin, serrated and as long as

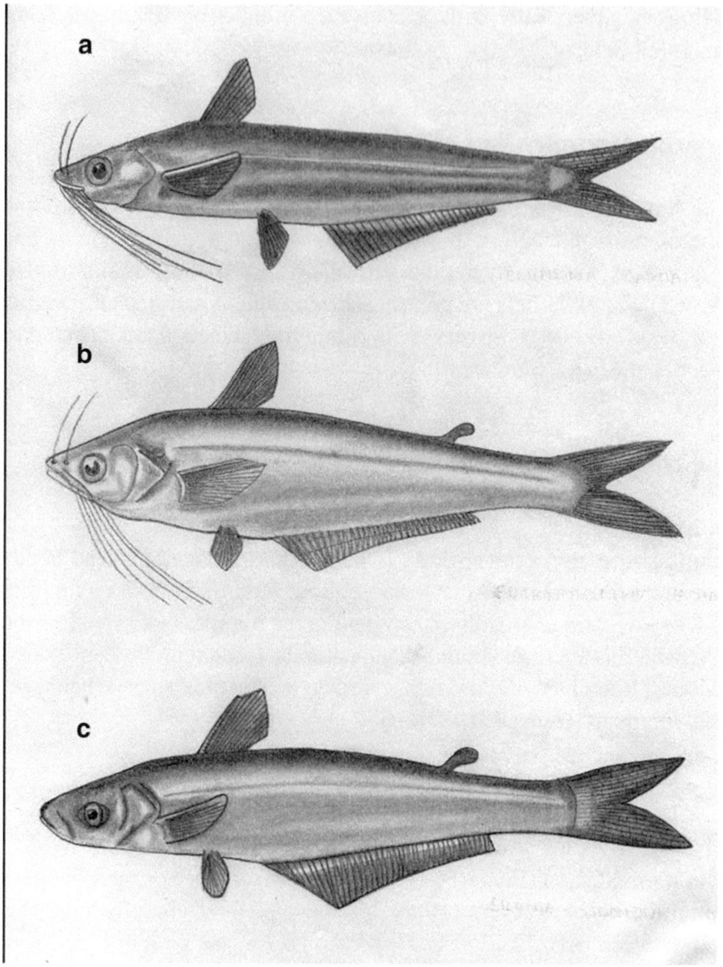

Fig. 3.5 Game fishes: (**a**) *Clupisoma garua*; (**b**) *Eutroplichthys vacha*; (**c**) *Silonia silondia*

3/4 of the length of head. Body is greyish, silvery and dark along the back (Day, 1878; Sehgal, 1987; Talwar & Jhingran, 1991).

1.18 *Silonia silondia* (Hamilton, 1822)

Common name: Silonid (Fig. 3.5).

Distinguishing characters: Body is elongate and compressed. Mouth is armed with rows of large and square sharp teeth. Maxillary pair of barbels are located in grooves not extending beyond eye. Mandibular barbels are very short. Caudal fin is large and strong. Dorsal fin is present with internally serrated spine. Anal fin is long.

Skin is smooth with body colour bluish above becoming silvery on sides and fins tipped grey (Gupta, 2015b; Rahman, 1989, 2005; Sehgal, 1987).

1.19 *Catla catla* (Hamilton, 1822)

Common name: Catla.

Distinguishing characters: Body is moderately compressed. Dorsal profile is more convex than the abdominal one. Eyes are with free orbital margin. Mouth is wide with a prominent lower jaw. Lower lip is large and folded. Dorsal fin is in advance of the ventrals. Pectorals extend to the ventrals. Body is greyish dorsally becoming silvery on the sides and beneath. Fins are of dark color becoming nearly black in some specimen (Sehgal, 1987; Talwar & Jhingran, 2001).

1.20 *Labeo calbasu* (Hamilton, 1822)

Common name: Kala Banse.

Distinguishing characters: It is a deep bodied fish with narrow mouth. Lateral lobe is absent but pores are present. Lips are thick fringed especially the lower one. Each lip is distinct with inner fold. Two pairs of barbels are present, the rostrals being slightly longer. Dorsal fin is midway between the snout and the base of the caudal fin. Caudal fin is deeply forked. The body is blackish with a scarlet centre in many of the scales (Bhuiyan, 1964; Rahman, 2005; Sehgal, 1987; Talwar & Jhingran, 2001).

1.21 *Labeo rohita* (Hamilton, 1822)

Common name: Rohu.

Distinguishing characters: It is deep bodied with a typical cyprinid shape, dorsal profile being more convex than the ventral. Lips are thick and fringed with a distinct inner fold above and below. Short and thin pair of maxillary barbels are present. Caudal fin is deeply forked, silvery on sides and beneath, brownish along the back with a red mark on each scale (Sehgal, 1987).

1.22 *Cirrhinus mrigala* (Hamilton, 1822)

Common name: Mrigal, Nain.

Distinguishing characters: The body is more or less cylindrical. Eyes are located on anterior half of the head. Two barbels are present. Dorsal fin reaches nearer to the snout than to the base of the caudal fin. Pectorals do not reach ventrals. Caudal fin is deeply forked. Body is silvery dark along the back, sometimes having coppery tinge and golden eyes. Pectorals, ventrals, and anal fins are orange colored stained with black (Sehgal, 1987).

1.23 *Pangasius pangasius* (Hamilton, 1822)

Common name: Pangas (Fig. 3.6).

Distinguishing characters: Upper jaw is larger than the lower. Barbels are four pairs, the maxillary pair extending upto the base of pectorals. Dorsal spine is serrated, moderately strong and almost equal to the length of pectoral spine. The color is silvery, darkest at the back with purple on sides, and under surface of head is golden (Day, 1878; Talwar & Jhingran, 1991).

1.24 *Salmostoma acinaces* (Valenciennes, 1844)

Common name: Silver razorbelly minnow (Fig. 3.6).

Distinguishing characters: Body is more or less compressed with a small head and upturned mouth. Cleft of mouth extends below the anterior third of the orbit. Sub-orbital ring of bones broad almost covers the cheek. Abdomen is not keeled. Dorsal fin is situated over the commencement of the anal. Caudal fin is deeply forked. Body is silvery with a lateral band (Rahman, 1989, 2005; Talwar & Jhingran, 1991).

1.25 *Bagarius bagarius* (Hamilton, 1822)

Common name: Goonch (Fig. 3.6).

Distinguishing characters: Body is elongated and compressed. Head is moderate and rounded. Eyes are small with a longer upper jaw. Dorsal spine is smooth, pectoral spine is stronger than and as long as dorsal spine. Caudal fin is deeply forked. All fins have a black base and often a dark band, greyish-yellow in color with irregular brown and black markings (Talwar & Jhingran, 1991).

2 Conclusion

Anglers have been enthralled with the variety of game fish species found in India's inland waters for many years. The nation's recreational fishing business greatly benefits from these fish, which range in size from the powerful Mahseer to the highly sought-after trout. They also provide an exciting angling experience. However, a number of anthropogenic and environmental factors are threatening the long-term viability of these game fish species. It is imperative that comprehensive conservation and management plans be put into place to guarantee that this beloved sport fishing heritage can continue to be enjoyed. This ought to entail actions like

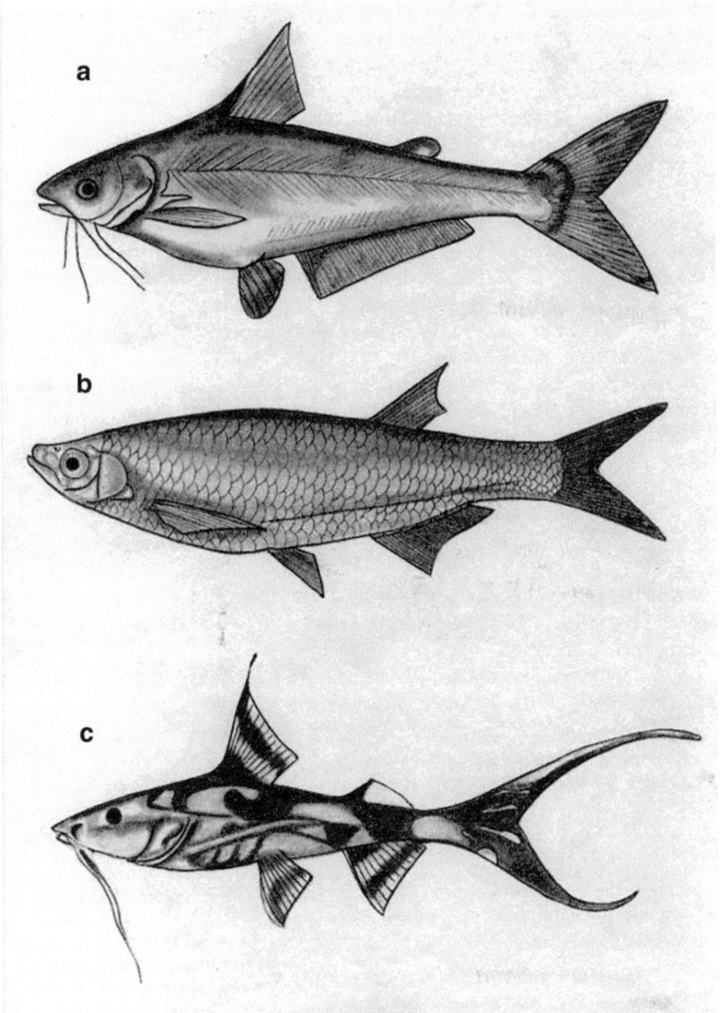

Fig. 3.6 Game fishes: (**a**) *Pangasius pangasius*; (**b**) *Salmostoma acinaces*; (**c**) *Bagarius bagarius*

protected area creation, pollution reduction, habitat restoration, and controlled fishing methods. Furthermore, additional study is required to comprehend these species' biology and ecology better so that more efficient management may be implemented.

References

Allen, G. (1991). *Field guide to the freshwater fishes of New Guinea* (p. 268). Madang.

Baishya, R. A., Basumatary, S., Kalita, H. K., Talukdar, B., Dutta, A., & Sarma, D. (2016). Present status and diversity of small indigenous fish species (SIS) in the upper reaches of river Brahmaputra in Assam, north-eastern India. *Indian Journal of Fisheries, 63*(1), 1–7.

Bhuiyan, A. L. (1964). *Fishes of Dacca* (p. 22). Aisiatic Society of Pakistan.

Campana, S. E. (1992). Measurement and interpretation of the microstructure of fish otoliths. *Canadian Special Publication of Fisheries and Aquatic Sciences, 117*(1), 59–71.

Courtenay, W. R., & Williams, J. D. (2004). *Snakeheads (Pisces, Channidae) - A biological synopsis and risk assessment* (p. 1251). US Geological Survey Circular.

Day, F. (1878). *Fishes of India, being a natural history of fishes known to inhabit the seas and freshwaters of India, Burma and Ceylon* (Vol. 1, p. 778). William Dawson & Sons Ltd..

Day, F. (1889). Fishes. In W. T. Blanford (Ed.), *The fauna of British India, including Ceylon and Burma* (Vol. I, pp. 1–548). Taylor and Francis.

Fuller, P. (2011). *Channa marulius. USGS nonindigenous aquatic species database*. US Geological Survey.

Gupta, S. (2015a). A review on Sperata aor (Hamilton, 1822), a popular food fish of Indian subcontinent. *Sains Malaysiana, 44*(9), 1217–1222.

Gupta, S. (2015b). Silonia silondia (Hamilton, 1822), A threatened fish of Indian subcontinent. *World Journal of Fish and Marine Science, 7*(5), 362–364.

Harrison, A. C. (1963). *Fresh-water fish and fishing in Africa*. Thomas Nelson and Sons.

Jhingran, V. G. (1959). Studies on the age and growth of Cirrhina mrigala (Ham.) from the river Ganga. *Proceedings of the National Academy of Sciences, India Section B: Biological Sciences, 25*(3), 107–137.

Kottelat, M. (2001). *Fishes of Laos* (p. 198). WHT Publications Ltd..

Kottelat, M., & Freyhof, J. (2007). *Handbook of European freshwater fishes* (p. 646). Publications Kottelat, Cornol and Freyhof.

Kullander, S. O., Fang, F., Delling, B., & Åhlander, E. (1999). The fishes of the Kashmir valley. In L. Nyman (Ed.), *River Jhelum, Kashmir Valley: Impacts on the aquatic environment* (pp. 99–167). SWEDMAR.

Laishram, I. S. (1991). *On the taxonomical description and distribution of copper Mahseer, Acrossocheilus hexagonolepis (McClelland)* (pp. 37–40). Records of the Zoological Survey of India.

Menon, A. G. K. (1999). Check list - Fresh water fishes of India. *Records of the Zoological Survey of India, 175*, 366.

Pant, B., Pandey, N. N., & Ram, R. N. (2018). Biometric characteristics based assessment of habitat suitability for the growth of a cold water cyprinid golden Mahseer (Tor putitora). *The Pharma Innovation Journal, 7*, 45–51.

Pethiyagoda, R. (1991). *Freshwater fishes of Sri Lanka* (p. 362). The Wildlife Heritage Trust of Sri Lanka.

Rahman, A. K. (1989). *Freshwater fishes of Bangladesh* (1st ed., p. 136). Zoological Society of Bangladesh.

Rahman, A. K. A. (2005). *Freshwater fishes of Bangladesh* (p. 115). The Zoological Society of Bangladesh.

Rainboth, W. J. (1996). *Fishes of the Cambodian Mekong. FAO species identification field guide for fishery purposes* (p. 265). FAO.

Sehgal, K. L. (1987). *Sport fisheries of India*. ICAR.

Shafi, M., & Quddus, M. M. (2001). *Fisheries of Bangladesh* (pp. 136–137). Kabir Publication.

Shahi, N., Sarma, D., Singh, B., Mallik, S. K., Baruah, D., Posti, R., Haldar, R., & Linthoingambi, I. (2023). Characterizing the dark Mahseer, Naziritor chelynoides (McClelland, 1839): A morphological, osteological, and molecular approach. *Genetics of Aquatic Organisms, 7*(2), 601.

Spillman, C.-J. (1961). *Faune de France: Poissons d'eau douce* (p. 303). Fédération Française des Sociétés Naturelles.

Taki, Y. (1974). *Fishes of the Lao Mekong basin* (p. 232). United States Agency for International Development Mission to Laos Agriculture Division.

Talwar, P. K., & Jhingran, A. G. (1991). *Inland fishes of India and adjacent countries* (Vol. 1, pp. 309–310). Oxford & IBH Publishing.

Talwar, P. K., & Jhingran, A. G. (2001). *Inland fishes of India and adjacent countries* (p. 203). Oxford and IBH Publishing.

Yamamoto, M. N., & Tagawa, A. W. (2000). *Hawai'i's native and exotic freshwater animals* (p. 200). Mutual Publishing.

Marine Game Fishes of India

4

Abstract

India's vast coastline and diverse marine ecosystems support a rich variety of game fish species that attract anglers from around the world. The most sought-after species include the Indo Pacific Tarpon (Megalops cyprinoides), Arabian Grouper (*Epinephelus tauvina)*, King Mackerel (*Scomberomorus guttatus*), and various species of croakers and billfish. These fish are prized for their acrobatic leaps, powerful runs, and impressive size, making them a thrilling challenge for recreational anglers. The best fishing locations are found along the Andaman and Nicobar Islands, Lakshadweep, and the southern coasts of Kerala and Tamil Nadu, where deep waters and nutrient-rich currents attract large pelagic predators. However, these game fish populations face threats from overfishing, habitat degradation, and the impacts of climate change. Effective management strategies, such as catch-and-release practices, size limits, and the establishment of marine protected areas, are crucial for ensuring the long-term sustainability of these valuable resources and the recreational fishing industry they support.

Keywords
Marine game fishes · Recreational fishing angling · Conservation · Sustainability

1 Introduction

Marine game fishes are fish species that are sought after by anglers for sport or recreation and are found in saltwater bodies such as oceans, seas, and gulfs. India is surrounded by the Indian Ocean on three sides, making it home to a diverse range of marine game fishes. These fishes are typically larger than their freshwater counterparts, with many species known for their size, strength, and fighting abilities. Some popular marine game fishes found in India include barramundi, tuna, sailfish,

© The Author(s), under exclusive license to Springer Nature Singapore Pte Ltd. 2024

S. T. Mushtaq et al., *Angling in India*,
https://doi.org/10.1007/978-981-97-8876-7_4

kingfish, and grouper. Anglers use a variety of techniques and equipment to catch these fishes, such as trolling, jigging, and live bait fishing. Many anglers practice catch-and-release fishing, which involves returning the fish to the water unharmed after catching them, to promote sustainability and conservation of fish populations.

In addition to being a popular sport, marine game fishes also serve as an important source of livelihood for coastal communities in India, with many fishermen relying on these species for their income. However, overfishing and habitat destruction pose significant threats to the sustainability of marine game fish populations, making conservation efforts critical for their survival.

1.1 *Megalops cyprinoides* (Broussonet, 1782)

Common name: The Indo-Pacific Tarpon, Oxeye herring (Fig. 4.1).

Distinguishing characters: It is a large silvery fish with bluish-green back, a very large mouth and a protruding lower jaw. It possesses a deeply forked tail, large scales and an elongate last dorsal-fin ray. The large eye is covered with adipose tissue and there is a conspicuous axillary scale at the base of each pectoral and pelvic fin. A bony gular plate is present between the jaw bones. Branchiostegals are more than 23 in number (Gomon et al., 2008; Sehgal, 1987).

1.2 *Epinephelus tauvina* (Forsskal, 1775)

Common name: Arabian Grouper or Greasy Rockcod (Fig. 4.1).

Distinguishing characters: Its body is elongated (depth 3.0–3.6 times in standard length); head large, (length 2.1–2.4 times in standard length). The interorbital area is narrow, flat to slightly concave, serrae at corner of preopercle slightly enlarged; upper edge of operculum almost straight; midlateral part of lower jaw with 2–5 rows of teeth. Rear nostrils distinctly larger than anterior nostrils. First gill arch with 8–10 gill rakers on upper limb, 17–20 on lower limb; no bony platelets on gill arch. Dorsal fin with 11 spines and 13–16 soft rays, the third to fifth spines longest, 3.1–4.7 times in head length and shorter than longest dorsal-fin rays; anal fin with 3 spines and 8 soft rays. Caudal fin is rounded. Pectoral-fin rays are 18 or 19 and lateral-line scales are 63–74 in number (Craig et al., 2011; Sehgal, 1987).

1.3 *Scomberomorus commerson* (Lacépède, 1800)

Common name: Narrow-barred Spanish Mackerel, King Seer (Fig. 4.1).

Distinguishing characters: Body is elongate (depth 4.8–5.6), moderately compressed; upper jaw reaching to posterior margin of eye or slightly beyond; teeth in jaws strong and compressed; gill rakers of first gill arch 0–2 on upper limb and 1–8 on lower limb, total 1–8. 2 dorsal fins are present and lateral line is abruptly bent

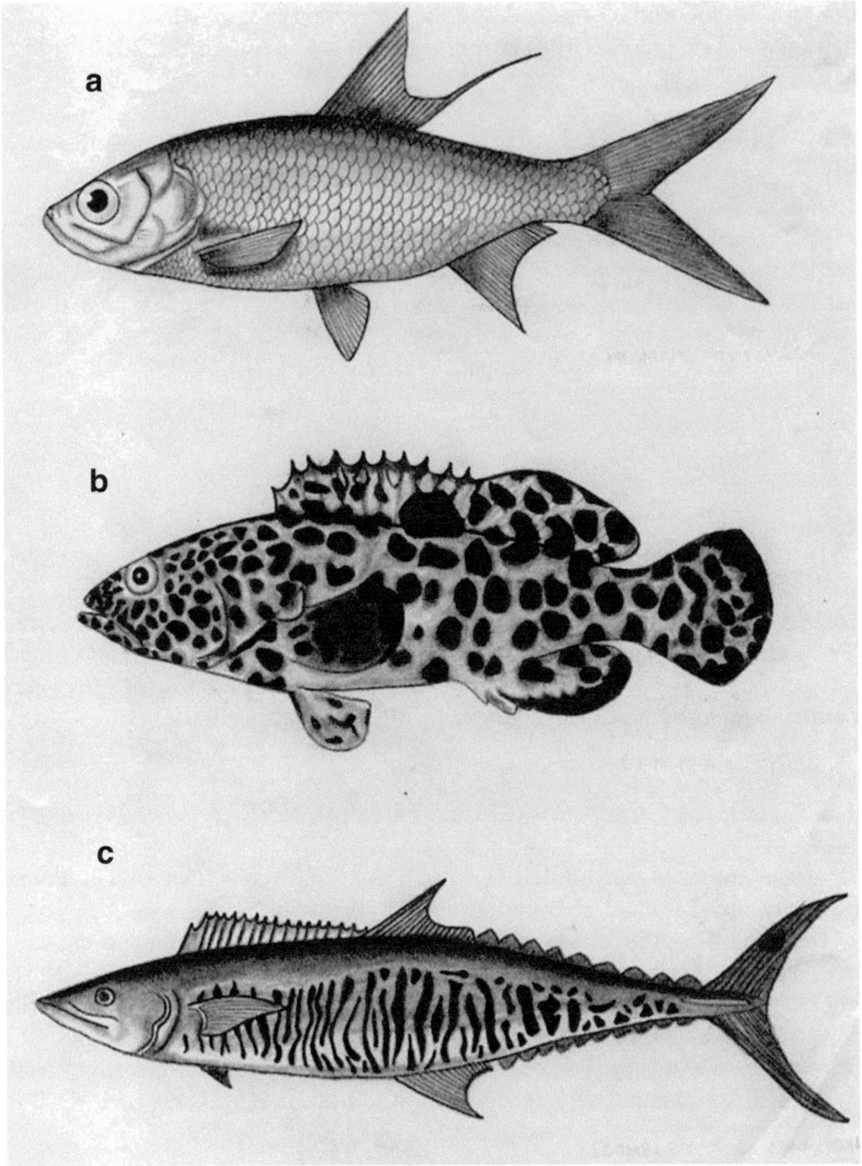

Fig. 4.1 Marine game fishes: (**a**) *Megalops cyprinoides*; (**b**) *Epinephelus tauvina*; (**c**) *Scomberomorus commerson*

downward below end of second dorsal fin. Large median keel with 2 smaller keels on either side are present (Allen & Erdmann, 2012; Collette, 2001; Sehgal, 1987).

1.4 *Scomberomorus guttatus* (Schneider, 1801)

Common name: Indo-Pacific King Mackerel.

Distinguishing characters: Body is torpedo-shaped. Maxilla reaching to hind border of eye. Length of head is 5–5.3 and body height is 5–5.2 times in total length. First dorsal fin possesses 16–17 weak spines and 8–9 finlets behind second dorsal. Anal fin is present with 3 weak spines and 7–9 finlets. Pelvics are small. Caudal peduncle is keeled (Collette & Nauen, 1983; Sehgal, 1987).

1.5 *Scomber albacares* (Bonnaterre, 1788)

Common name: Yellow Fin Tuna, Longfin Tuna.

Distinguishing characters: There is a large median keel with two smaller keels on either side of caudal peduncle. Pectoral fins are moderate in length with bluntly pointed tips. Second dorsal fin and anal fin is yellow, more elongated in adults. 26–34 gill rakers on first gill arch. 10 dorsal finlets and 7–10 anal finlets are present. Interpelvic process with two flaps are present. A distinct notch is seen in central portion of trailing edge of caudal fin (Sehgal, 1987: Collette, 2001).

1.6 *Lutjanus argentimaculatus* (Forsskal, 1775)

Common name: Mangrove Red Snapper, Creek Red Bream, Dog Bream, Purple Sea Perch, River Roman, or Rock Barramundi (Fig. 4.2).

Distinguishing characters: It has a slightly concave caudal fin and the scale rows on the back are roughly parallel to the lateral line. The first gill arch has 16–20 gill rakers with the lower limb having 9–12 (including rudiments). The dorsal fin has 10 spines and 13–14 soft rays, the anal fin has 3 spines and 8 soft rays. The posterior profile of dorsal and anal fins is rounded. The pectoral fins have 16–17 rays (Allen & Erdmann, 2012; Anderson & Allen, 2001; Sehgal, 1987).

1.7 *Protonibea diacanthus* (Lacepède, 1802)

Common name: Blackspotted Croaker (Fig. 4.2).

Distinguishing characters: The snout is pointed with a large terminal mouth forming a low angle to the horizontal. Teeth are differentiated into large and small in both jaws. No canine teeth present. Lower gill rakers are 7–8 in number. Dorsal fin spines are 9–10 followed by low notch and 22–24 soft rays. Pectoral fin is fairly small, a little more than half of the head. Anal fin is with two spines and seven soft

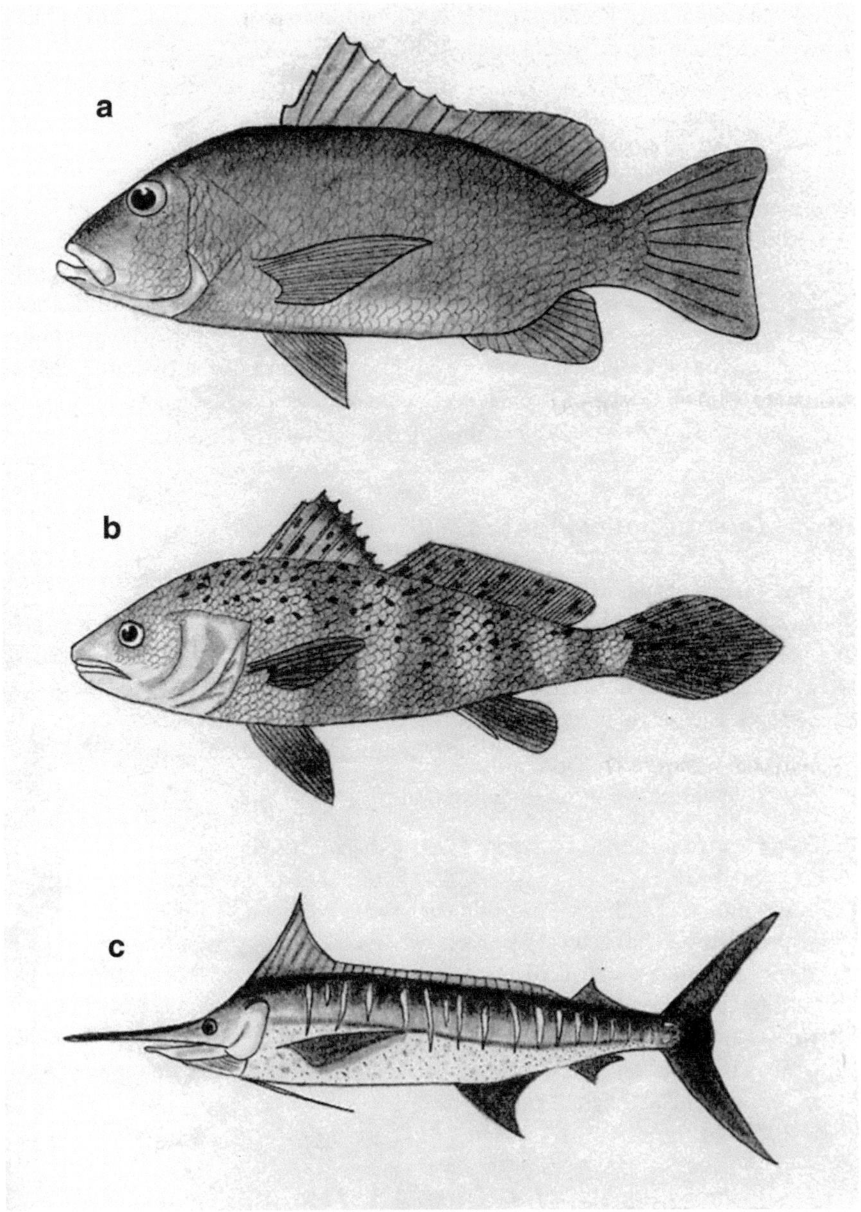

Fig. 4.2 Marine game fishes; (**a**) *Lutjanus argentimaculatus*; (**b**) Protonibea diacanthus; (**c**) Makaira nigricans

rays. The second anal spine is strong. Rhomboid shaped caudal fin. Cycloid scales are present on snout and eyes, rest of the scales are ctenoid. Five dark blotches are present along back with many small black spots on top of head, upper half of body, dorsal and caudal fins. Pectoral, pelvic, anal, and lower part of caudal fins is black in colour (Kizhakudan & Kizhakudan, 2017; Sehgal, 1987).

1.8 *Makaira nigricans* (Lacépède, 1802)

Common name: Blue marlin, Atlantic Blue Marlin, Billfish (Fig. 4.2).

Distinguishing characters: The body is cylindrical from anal fin forward. Two dorsal fins are present; the first dorsal fin is high and slopes steeply posteriorly, while the second fin is small. The caudal peduncle has keels. The lateral line forms a large net-like pattern of hexagons canvasing the sides of the fish. The pelvic fins are slender. 7–8 gill rakers are present on the lower limb of the first arch. Dorsal spines are 10–11 in number. Dorsal soft rays are 225 in number. Two anal spines and seven anal soft rays are present (Sehgal, 1987; Smith, 1997).

1.9 *Lates calcarifer* (Bloch, 1790)

Common name: Bhetki, Asian Sea Bass, Baramundi (Fig. 4.3).

Distinguishing characters: Body is humped-back. Mouth is armed with minute file-like teeth. Maxilla extends beyond eye. Length of head is 3.6–4 and body height is 3.5–3.6 times in total length. Two dorsals are united at their bases. Anal fin has three spines. Caudal fin is round. Color is grey with green above, silvery below during the monsoons with a tinge of purple (Froese & Pauly, 2003; Sehgal, 1987).

1.10 *Eleutheronema tetradactylus* (Shaw, 1804)

Common name: Fourfinger Threadfin (Fig. 4.3).

Distinguishing characters: Mouth is set below with upper portion carrying the nose; very prominent with bold large eyes. Mouth is armed with minute file-like teeth. Maxilla reaches beyond the eye. Length of head is 5–5.2 and body height is 5–6 times in total length. Pectorals have four free rays extending to pelvics. Distance between pelvic and anal bases equals to 3/4 length of head (Motomura, 2004; Motomura et al., 2002; Sehgal, 1987).

1.11 *Acanthopagrus berda* (Forskal, 1775)

Common name: Goldsilk Seabream, Sly Bream or Picnic Seabream (Fig. 4.3)

Distinguishing characters: Dorsal profile is more convex than ventral. Snout is pointed. Length of head is 4–4.2 and body height is 2.5–2.6 times in total length. It has a single dorsal fin with 11–12 rigid normal spines. Three anal spines are present, with second being the largest. The pelvis does not reach the anus. Caudal fin is emarginated (Iwatsuki & Heemstra, 2010; Sehgal, 1987).

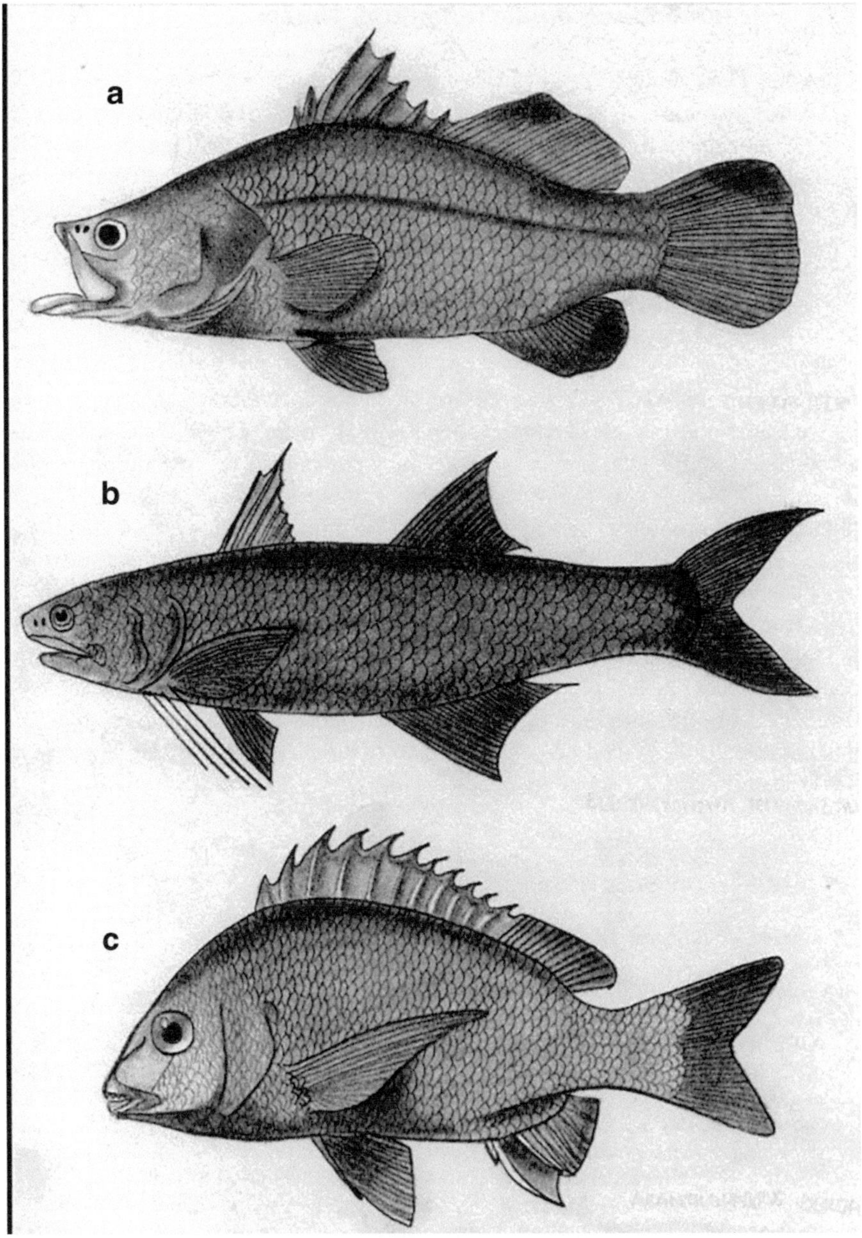

Fig. 4.3 Marine game fishes: (**a**) *Lates calcarifer*; (**b**) Eleutheronema tetradactylus; (**c**) Acanthopagrus berda

1.12 *Acanthopagrus datnia* (Hamilton, 1822)

Common name: Bengal Yellowfin Seabream.

Distinguishing characters: Dorsal profile is much elevated. Snout is pointed. Length of head is 4–4.2 and body height is 2.5–2.7 times in total length. Dorsal fin is single with 11–13 rigid normal spines, the spiny part longer than soft part. Pelvics do not reach anus. Pectorals are longer than the head length. Caudal fin is emarginated (Hasan et al., 2020; Sehgal, 1987).

2 Conclusion

India's marine game fish resources are an important resource that support the nation's coastal economies and recreational fishing sector. These iconic species are essential to keeping marine ecosystems healthy and in balance, in addition to offering exciting fishing opportunities. However, a number of stressors brought on by humans and alterations in the environment pose a threat to the long-term viability of these game fish species. Comprehensive conservation and management measures must be put into place in order to guarantee that future generations can continue to experience the thrill and challenge of following these amazing creatures. In order to maintain sustainable harvest levels, this should entail taking steps like creating marine protected zones, putting catch-and-release procedures into place, and controlling fishing activity. Furthermore, additional study is required to comprehend these species biology and ecology better so that better management choices can be made.

References

Allen, G. R., & Erdmann, M. V. (2012). *Reef fishes of the East Indies* (Vol. 1). Universitiy of Hawai'i Press.

Anderson, W. D., & Allen, G. R. (2001). Lutjanidae. Jobfishes. In K. E. Carpenter & V. Niem (Eds.), *FAO species identification guide for fishery purposes. The living marine resources of the Western Central Pacific. vol. 5. Bony fishes part 3 (Menidae to Pomacentridae)* (pp. 2840–2918). FAO.

Collette, B. B. (2001). Scombridae. Tunas (also, albacore, bonitos, mackerels, seerfishes, and wahoo). In K. E. Carpenter & V. Niem (Eds.), *FAO species identification guide for fishery purposes. The living marine resources of the Western Central Pacific* (Vol. 6. Bony fishes part 4 (Labridae to Latimeriidae), estuarine crocodiles, pp. 3721–3756). FAO.

Collette, B. B., & Nauen, C. E. (1983). *FAO species catalogue. Vol. 2. Scombrids of the world. An annotated and illustrated catalogue of tunas, mackerels, bonitos and related species known to date* (p. 137). FAO.

Craig, M. T., Sadovy de Mitcheson, Y. J., & Heemstra, P. C. (2011). *Groupers of the world: a field and market guide* (p. 356). CRC Press.

Froese, R., & Pauly, D. (2003). Dynamics of overfishing. In J. L. Lozán, E. Rachor, J. Sündermann, & H. von Westernhagen (Eds.), *Warnsignale aus Nordsee und Wattenmeer-eine aktuelle Umweltbilanz* (pp. 288–295). GEO.

Gomon, M., Bray, D., & Kuiter, R. (2008). *Fishes of Australia's southern coast* (p. 928).

Hasan, M. E., Durand, J. D., & Iwatsuki, Y. (2020). Acanthopagrus datnia (Hamilton. 1822), a senior synonym of Acanthopagrus longispinnis (Valenciennes, 1830) (Perciformes: Snaridael). *Zootaxa, 4750*, 151–181.

Iwatsuki, Y., & Heemstra, P. C. (2010). Taxonomic review of the Western Indian ocean species of the genus Acanthopagrus Peters, 1855 (Perciformes: Sparidae), with description of a new species from Oman. *Copeia, 2010*, 123–136.

Kizhakudan, J. K., & Kizhakudan, S. J. (2017). *Prioritized species for mariculture in India* (pp. 255–259). ICAR-Central Marine Fisheries Research Institute.

Motomura, H. (2004). *Threadfins of the world (Family Polynemidae): An annotated and illustrated catalogue of polynemid species known to date (No. 3)*. FAO.

Motomura, H., Iwatsuki, Y., Kimura, S., & Yoshino, T. (2002). Revision of the Indo-West Pacific polynemid fish genus Eleutheronema (Teleostei: Perciformes). *Ichthyological Research, 49*, 47–61.

Sehgal, K. L. (1987). *Sport fisheries of India*. ICAR.

Smith, C. L. (1997). *National Audubon Society field guide to tropical marine fishes of the Caribbean, the Gulf of Mexico, Florida, the Bahamas, and Bermuda* (p. 720). Alfred A. Knopf, Inc.

Sport Fishing Sites in India

5

Abstract

From the tropical coasts to the Himalayan rivers, India's varied landscapes provide a plethora of options for those who enjoy sport fishing. Anglers from all over the world are drawn to the Andaman and Nicobar Islands because of their pristine reef ecosystems and abundance of pelagic game fish, including marlin, tuna, and sailfish. Rivers such as the Beas, Indus, and Lidder in the Himalayan region are renowned for their populations of valued Mahseer and Trout, offering exciting angling experiences. Locations such as Kovalam and the Coromandel Coast on the southern shores provide access to species such as Barracuda, King Mackerel, and Yellowfin Tuna. The Ramganga River in Corbett National Park and other inland lakes are home to a variety of game fish, including Mahseer and Goonch. With extensive rules, knowledgeable guides, and a burgeoning recreational fishing sector, India is becoming a top sport fishing destination where anglers may pursue a variety of iconic species in breathtaking natural settings.

Keywords

Sport fishing · Game fish · Mahseer · Trout · Himalayan rivers · Coastal waters

1 Introduction

India has a rich diversity of freshwater and marine ecosystems that offer a wide range of sport fishing opportunities. Angling or sport fishing has remained a favorite pursuit of the Britishers who introduced brown trout and rainbow trout in the upland waters of the country and has now become a very popular activity giving boost to tourism especially in the Himalayas. Angling is the chief technique of sport fishing. However some angling techniques such as long lining or trolling are also used in commercial fisheries. Other techniques such as catch and release fishing is mostly

© The Author(s), under exclusive license to Springer Nature Singapore Pte Ltd. 2024

S. T. Mushtaq et al., *Angling in India*,
https://doi.org/10.1007/978-981-97-8876-7_5

used by amateur fishermen. Some of the famous sites of India where both domestic and international anglers seek adventure holidays are listed below.

1.1 Jammu and Kashmir

The valley of Jammu and Kashmir is considered to be an angler's paradise as it is blessed with several snow and spring fed streams. It is located in the northernmost part of India and is known for its breathtaking natural beauty, serene lakes, and rivers. The region is also a popular angling destination, attracting fishing enthusiasts from all over the world. Angling in various pristine lakes of Kashmir like the Dal Lake and Nigeen Lake offers an everlasting experience. The streams of Kashmir are full of Brown and Rainbow Trout fishes. The best season for fishing in the state is from April to October. The Sindh, Kishen Ganga (Fig. 5.1), Doodh Ganga, and Lidder rivers are generally covered with snow and begin to melt between the months of May to July. These rivers are well stocked with fishes and offer an everlasting experience to the tourists. There are some alpine lakes such as Nilnag as well, which offer an incredible experience of trout fishing. Angling in Kashmir is regulated by the Jammu and Kashmir Fisheries Department which issues permits for fishing in specific water bodies. The department also imposes strict regulations on the size and quantity of fish that can be caught. Around 61 beats are open for fishing and angling in Kashmir during the season. Each beat is about 3 to 4 kms long and is connected with a motorable road except Krishensar, Vishensar, and Gangabal lakes (Fig. 5.2), where one has to trek to reach the destination (Table 5.1; Fig. 5.3). Fishing enthusiasts visiting Kashmir can hire experienced guides who can help them navigate the local regulations and find the best spots for angling. The guides can also provide equipment and gear making the entire experience hassle free. Before one starts the angling exercise, it is mandatory that to get a proper license from the Directorate of the Fisheries, Srinagar. No angler is allowed to catch more than six fishes. One also has to keep in mind that only fly fishing is permitted in Kashmir, while other methods are strictly restricted. After catching the fish, the tourists are required to release them back into the water. All these measures have been taken by the state government to preserve and maintain the environmental balance.

1.2 Uttarakhand

This state is located in northern India and is a popular sport fishing destination. The state is home to several rivers, lakes, and streams that offer ample opportunities for anglers to catch a variety of fish species. Corbett National Park nestled in Nainital district of Uttarakhand is a famous destination for the fishing and angling enthusiasts. The park houses many exotic varieties of flora and fauna and the pristine Ramganga River flowing by the park is an abode of several fish species such as the Mahseer, Goonch, Brown Trout, and many other kinds of fishes. However, the permission of the Divisional Forest Officer (DFO) of the park is required for fishing

Fig. 5.1 Tulail valley, Kishenganga Beat, Gurez river, J&K. Picture credit: Anjuman Sameen

and angling in Corbett National Park. The private resorts located in close proximity to the park can arrange the fishing permit license and angling trips. Fishing in Ramganga can be carried out all round the year. However, October to mid- June is considered to be the best season to indulge in the recreational activity of angling and fishing. It is important to note that fishing and angling is permitted in the Upper Ramganga, extending from Nagataley to Masi. Bhagirathi river, a tributary of the Ganga is known for its trout population and also finds its place in some of the best sport fishing sites in India. Yet another important river famous for Mahseer and also as a sport fishing site is Kali river located on Indo-Nepal border. The river originates in the watershed bordering India and Tibet. By the time it meets the Gori Ganga, it is well out of the alpine zone and into the verdant hilly landscape of Kumaon. The

Fig. 5.2 Fly fishing at Gangabal Lake. Picture credit: Dr. Musadiq Allaqaband

best spot to fish at is Pancheshwar, which is a beautiful village located about 40 km from Lohaghat town in Champawat where the Kali meets the Saryu river. The water here is crystal clear and full of many kinds of fish, especially the Golden Mahseer. Besides the Golden Mahseer, Pancheshwar may also yield Copper Mahseer, or the rare Red fin Mahseer. The best season for fishing is before the arrival of the rains and after the month of October. People from various geographical locations of the world flock to this place to participate in the International Angling Competition held in March every year.

Table 5.1 Trout beats in Jammu and Kashmir (DOF, JK)

S. No.	Trout beats	Stream/river	District
1.	Wayil, lower, Wossan, middle, Kangan, Cherwan, Mammer, Sumbal, Gund, Rezin	Sindh	Ganderbal
2.	Lower, middle, upper; Krishensar-Vishensar; Gangabal, Satsar, Gadsar; Sonamarg Rec. Park; Wangdara trout stream	Wangath	
3.	Lower, middle, upper	Erin	Bandipora
4.	Shera Sar; Madhumati upper, middle, lower	Madhumati	
5.	Kishenganga	Gurez river	
6.	Hamal upper, Hamal middle; Bijhama trout stream; Gulmarg Rec. park; Ningle trout stream; Barzil Nallah; Lachi Para stream	Hamal trout stream	Baramulla
7.	Lower, middle, upper; Boniyar trout stream	Ferozepur	
8.	Lower, middle, upper	Naristan trout stream	Pulwama
9.	Upper, middle	Lam stream	
10.	Aripal trout stream	Aripal trout stream	
11.	Romshi Nallah	Romshi Nallah	
12.	Lower, middle, upper	Doodganga stream	Budgam
13.	Lower, middle, upper	Sukhnag trout stream	
14.	Shaliganga trout stream	Shaliganga trout stream	
15.	Lower, middle, upper	Aharbal trout stream	Kulgam
16.	Wasak Nag	Wasak Nag	
17.	Kausar Nag	Kausar Nag	
18.	Lower, middle, upper	Hirpora trout stream	Shopian

(continued)

Table 5.1 (continued)

S. No.	Trout beats	Stream/river	District
19.	Lower, middle, upper	Mawar trout stream	Kupwara
20.	Dringyari trout stream	Dringyari trout stream	
21.	Kalaroos trout stream	Kalaroos trout stream	
22.	Nafran trout stream	Lidder, Pahalgam sector	Anantnag
23.	Aru stream	Aru stream	
24.	Aru lower, upper; Seeshnag; Phrysian; Masjid; Langanbal; Batkote; Yenner; Bumzoo; Sakras; lower; Wahdhan upper, lower; Tricker stream (1–6); Nambal; Kotsu; Langnai stream; Panzeth stream; Chattergul stream; Achabal stream; Kausar nag; Sheeshnag sar	Lidder, Pahalgam sector	
25.	Lower, middle, upper; Dessu; Bidhar; Daksum; Mather	Bringhi (Daksum)	
26.	Lower, middle, upper	Nowbugh stream	
27.	Upper, lower	Ahlan stream	
28.	Badhora stream	Badhora stream (Mohre)	Reasi
29.	Sui devta stream	Sui Devta stream	
30.	Kirchi stream	Kirchi stream	Udhampur
31.	Dhaggar Nallah	Dhaggar Nallah	Kathua
32.	Marwah Wardwan stream	Marwah Wardwan stream	Kishtwar
33.	Singapore stream	Singapore stream	
34.	Famber stream	Famber stream	
35.	Keshwan Nallah	Keshwan Nallah	
36.	Mahu-mangat stream	Mahu-mangat stream	Ramban
37.	Khet Nallah	Khet Nallah	
38.	Bheja stream	Bheja stream	Doda
39.	Haloon Nallah	Haloon Nallah	
40.	Jai Nallah	Jai Nallah	

(continued)

Table 5.1 (continued)

S. No.	Trout beats	Stream/river	District
41.	Ans Nallah	Ans Nallah	Rajouri
42.	Gabbar stream	Gabbar stream	
43.	Bufliaz Nallah	Bufliaz Nallah	Poonch
44.	Loran Nallah	Loran Nallah	

Fig. 5.3 Aharbal beat. Picture credit: Dr. Musadiq Allaqaband

1.3 Himachal Pradesh

Himachal Pradesh is an attractive destination for fishing and angling. The Pabbar Valley region, located around 80 kms, north-east of Shimla is one of the best place for Golden Mahseer Fishing in India. Here fishing is carried out on the banks of Pabbar River (Fig. 5.4). The other popular places for fishing and angling in Himachal Pradesh are Mandil, Dhamvari, Chirgaon, Seema, and Sandasu. The Beas, Sutlej, Ravi, and Baspa flow through different districts of Himachal Pradesh, mainly in the upper Himalayas. These rivers are goldmine for brown Trout fishing. Tributaries of the Beas River like Sarvari, Parbati, Sainj, Hurla, Phojal, and Tirthan are considered excellent destinations for fishing. Here, the fishing enthusiasts can get a wonderful

Fig. 5.4 Pabbar river. Snjsharma, CC BY-SA 3.0. https://creativecommons.org/licenses/by-sa/3.0. Via Wikimedia Commons

chance to catch various kinds of fishes like the Brown Trout, catfish, Rohu, Catla, Golden Masheer, Mrigal, *Nemacheilus* spp., *Barilius* spp., and Schizothoracids.

1.4 Arunachal Pradesh

Angling and fishing in Arunachal Pradesh offers a great chance to get immersed in the beauty of nature. There are four prominent regions in the state that are popular for fishing. These regions include Bhalukpong and Tipi, located on the banks of Kameng River; Pasighat and Wakro situated on the banks of Siang and Lohit River. As the rivers are translucent, there is high oxygen content in these rivers and hence there is a high possibility of a good catch. The main fishes found in the rivers are Golden Mahseer and Brown Trout. It is the best location for brown trout fishing. The angling sports of Arunachal are very peaceful and offer a magnificent opportunity for fast and still water angling.

Not many anglers who have pursued the Golden/Chocolate Mahseer fish in India can claim to have made a cast in the largely unexplored waters of the remote Dibang River in the northeastern state of Arunachal Pradesh. The river originates in the snow-covered southern flank of the Himalayas, close to the Indo-Chinese border, and flows southwards to meet the mighty Brahmaputra. Along the way, it is fed by as many as a dozen tributaries, big and small. The abundance of water coupled with lack of enough fishing opportunities in the past have made the Dibang river an ideal site for the anglers who want to traverse the untraveled territory and fish the unexplored river in search of the Golden and the Chocolate Mahseer.

1.5 Assam

The north-eastern state of Assam is a paradise for angling and fishing enthusiasts. The Jia Bhoroli River in Assam that emanates in Tawang district of Arunachal Pradesh and flows through Sonitpur district of Assam is famous for Golden Mahseer fishing. An annual angling competition is organized in the month of November every year in association with the State Forest Department of Assam and Assam Bhoroli Anglers Association. Tourists from India and other parts of the world swarm this place to participate in the competition. Apart from the competition, the months of October to April are the best time for fishing. The other kinds of fishes found in the Jia Bhoroli River are Saal, Gorua, Korang, and Boka. Kapili, Manas, and Bhalukpung rivers are also ideal spots for angling.

1.6 Meghalaya

Meghalaya has a picturesque beauty and exotic wildlife. However, besides its wildlife, it is also becoming one of the popular destinations for angling and fishing trips in India. Ranikor, a small town, nestled about 5 hours from Shillong, the capital of Meghalaya, is a famous destination for fishing. The Kynshi River that originates from Rilang town is a rich store of various species of fishes such as the Golden Mahseer, Chocolate Mahseer, Goonch (Catfish), and Eels. The tribals can also be seen catching fishes using traditional equipments. A fishing trip to Meghalaya is really a captivating experience, where the fishes are caught by riding on a local boat.

1.7 Karnataka

Fishing and angling is one of the unique recreational activities carried out by the people in Karnataka. The state government runs angling camps across the Cauvery River, which houses several varieties of Mahseer. It is very interesting to know that the Mahseer are considered to be the "Pride of Cauvery River" and various concrete steps have been undertaken to safeguard the their habitat. In the fishing and angling camps, generally run by the government, catch and release practice is followed. Every time a fish has been hooked, the camp attendant helps the visitors to remove the fish from the water, tie the mouth by using a nylon rope, take the weight, and pose for a photograph. After all these activities, the fish is then released into the water.

1.8 Kerala

When one envisages about Mahseer fishing and angling trip to South India, then the exotic state of Kerala immediately comes to the mind. The beguiling streams of Munnar hill station and Elephant Lake provide a pleasurable fishing and angling experience. The streams contain a stockpile of Trout fishes, Mahseer, etc., and catching a fish amidst the verdant landscape provides memories of a lifetime. The best time for fishing in Kerala is from October to March. Silver, Gold, Black, and Pink Mahseer, Pearl Spot, Climbing Perch, Sardine, Pomfret, Tuna, Seer, and Mackerel are the other varieties of fishes found in the rivers and streams of Kerala.

1.9 Goa

Goa is one of the best angling destinations in India that offers a great chance to the angling lovers to trap various kinds of fishes in the tropical surroundings of the state. Many beaches like the Agonda beach, Vagator beach, Chapora River, etc., contain several varieties of fishes like the Soormai, Salmon, Mullet, Ladyfish or catfish, Rockfish, Stingray, Guitarfish, the local Barramundi, and many more. The ideal time for fishing in Goa is October to May. Many visitors patiently wait by the beaches to get their prized catch and pose pictures with it.

1.10 Andaman And Nicobar Islands

The Andaman and Nicobar Islands are enveloped by blue waters that offer a great chance for fishing and angling. The islands with its lush rainforests and mangrove trees certainly mesmerize the tourists. The rivers of Andaman are a home to various kinds of fishes like Dorado, Black Marlin, Yellowfin Tuna, and Dogtooth. Saltwater fishing is famous among the tourists in the Havelock Islands of Andaman.

1.11 Lakshadweep

Lakshadweep is also a group of islands that not only charms the visitors with its serene beauty but also with its magnificent beaches. The beaches are the perfect avenue for fishing and angling. The waters of the beaches are very shallow and transparent. The anglers can look forward to catch various kinds of fishes like the Dolphin fish, Flying fish, Yellowfin, Rainbow Runner, Red Snapper, Butterfly fish, Skip Jack, Sail fish, Parrot fish, and Coral fish. Besides, fishing, there are also many enthralling adventure activities that can be enjoyed in this beautiful island like surfing, kayaking, scuba diving, and canoeing.

2 Conclusion

Sport fishers have a plethora of options thanks to India's varied habitats and terrain. India offers both domestic and foreign anglers an exhilarating angling experience, from the immaculate lakes and rivers of Jammu and Kashmir to the Himalayan rivers of Uttarakhand and the tropical shores of Andaman and Nicobar Islands. To maintain these priceless resources, effective conservation and management techniques are essential. The long-term sustainability of these game fish populations depends on policies like size restrictions, catch-and-release procedures, and the creation of marine protected zones, which are all necessary to preserve the ecological equilibrium. India can continue to draw anglers from all over the world, boosting tourism and bolstering local economies, by encouraging sustainable sport fishing techniques and protecting these iconic species. India can set the standard for responsible and sustainable sport fishing methods by collaborating with scholars, legislators, conservation organizations, and fishermen.

Sport Fisheries Potential in the Himalayan Region

6

Abstract

Recreational fishing for Mahseer, trout, and snow-trout is a popular activity among affluent families and tourists in India, particularly in the Himalayan regions. The physiography and productivity of these cold-water ecosystems indicate significant potential for developing sport fisheries. The Himalayan uplands, with their diverse streams, rivers, and lakes, are home to both exotic trout species and native Mahseer, making them attractive destinations for anglers. Notably, Jammu and Kashmir, Himachal Pradesh, Uttarakhand, and the Northeast region of India offer abundant fishing opportunities with various streams, rivers, and lakes supporting sizeable catches. The development of sport fisheries in these areas could bolster local economies, create employment, and enhance adventure tourism. Effective strategies, including conservation plans, infrastructure development, artificial propagation, and strict fisheries regulations, are essential to harness this potential sustainably.

Keywords

Recreational fishing · Mahseer · trout · Snow-trout · Himalayan rivers · Cold-water ecosystems · conservation

1 Introduction

In India, the sport or recreational fishing for Mahseer, trout, and snow-trout is indulged in by a section of relatively well to do families who are fond of sports and outdoor activities. It is also cherished by the tourists visiting India for adventure and enjoyment. Recreational angling has both social and economic benefits worldwide, and these benefits may be large in underdeveloped nations. From the information available on physiography and productivity status of main cold-water

© The Author(s), under exclusive license to Springer Nature Singapore Pte Ltd. 2024

S. T. Mushtaq et al., *Angling in India*,
https://doi.org/10.1007/978-981-97-8876-7_6

ecosystems in this region, it has been revealed that there is ample potential to develop sport fishery of both exotic trouts and native Mahseer species in the Himalayan uplands (streams, rivers, and lakes). Because trout and mahseer fishing appeals to a wide range of tastes and pursuits, it has become a popular pastime for tourists drawn to India's northern and southern uplands. Most of the glacial rivers in their upper reaches and all snow-fed streams are home to the exotic trouts, especially the brown trout.

The important trout waters in the Himalayan region are most of the snow-fed streams including Lidder, Simal, Erin, Madhumati, Tricker, Kokernag, Verinag (J&K), upper reaches of Beas and Sutlej (Himachal Pradesh), Ashiganga, Birehi, Pinder, Bhagirathi (Uttarakhand) besides a few upland cold-water lakes/streams in North-East region. The lower stretches of these upland streams/rivers meandering in lesser or outer Himalayas are good resources of Golden Mahseer. The frequent catches of sizeable Mahseers by anglers in these areas support the views to develop them as Mahseer waters for game fishing purposes. Among the Himalayan streams/rivers, Beas and its tributaries in the foot hill region and rivers, Giri (HP), Yamuna between Tajwala (Haryana) to Dhak Pathar (Uttaranchal), Ganga between Rishikesh to Tehri and its tributaries, the Kali, Saryu, East and Western Ramganga, East and Western Nayar, Song, Kosi (Uttarakhand), Chenab and its tributaries, Tawi and Anji (J&K), and Rivers Jia-Bhoreli, Diwang, Subansini, and Manas (NEH Region) are important as Mahseer fishery resources. Apart from these, the lakes of Kumaon (Bhimtal, Khurpatal, Naukuchiatal, Nal-Damyantital, and Sattal do contribute substantially to Mahseer fishery and provide ample scope for fish sport.

Rainbow trout is native to the cold water rivers and lakes of North America's and Asia's Pacific coasts. Because rainbow trout tolerates a wider range of environmental and production circumstances than other trout species, it has been introduced to around 82 countries almost everywhere the conditions are appropriate for its culture. In 1877, the first successful shipment of rainbow trout outside of North America occurred (Canyurt & Akhan, 2009). Brown trout, *Salmo trutta fario* and rainbow trout, *Oncorhynchus mykiss* were introduced to India from Europe largely to promote sport fishing or recreational angling (Jhingran & Sehgal, 1978).

The introduction of brown trout (*Salmo trutta*) and rainbow trout (*Oncorhynchus mykiss*) by British anglers in streams and rivers of the Himalayas and Western Ghats in the 1860s and 1909, respectively, aided to increase the popularity of this leisure pastime. Trout fish were first documented in Himachal Pradesh in Kangra and Kullu Districts in 1905 from Jammu and Kashmir (Sehgal, 1974). Trout farming in Himachal Pradesh began in 1909 with the shipping of eyed ova from Srinagar to Kullu Valley. Until 1980, the culture was limited to producing fish fingerlings and placing them in rivers and streams. Trout was only regarded a game fish, and there was little interest in its culture or large-scale production (Sehgal, 1999).

2 Potential of Sport Fishery in Jammu and Kashmir

The Union territory of Jammu and Kashmir has a great potential to become one of the best adventure tourism destinations in the world. It offers wonderful exotic Himalayan ranges and landscapes, fast flowing cold-water rivers, streams, and lakes. Sport fishing is an important part of adventure tourism and the valley is called as anglers paradise as it provides many opportunities for anglers with its varied types of fishes especially trout and Mahseer (Figs. 6.1 and 6.2). Recreational fisheries is an important component of the adventure tourism industry and also important for local and regional development around the state. The valley has many promising angling sites. There is an "ideal" fishing permit that permits you to fish only in a

Fig. 6.1 Trout captured by angler Ubaid Wani from Kashmir mountain stream

Fig. 6.2 Angler Sameer with Mahseer catch at Basoli Dam, Jammu. Picture Credit: Dr. Musadiq Allaqaband

2-km beat (Mandal et al., 2018). The sport of angling can further become a popular recreational activity/hobby among the locals, tourists, and professional anglers provided the government and stakeholders adopt a well-planned strategy for the development of this sector.

3 Potential of Sport Fishery in Himachal Pradesh

Himachal Pradesh has some of the finest trout streams in the north. Some of these include the Pabbar in the Rohru valley, the Baspa in the Sangla valley, the Uhl in the Barot valley, and river Beas and its tributaries in the Kullu valley which are rich in both brown and rainbow trout, while many rivers and streams in the Kangra valley are well-known for mahseer. Each of these rivers has 32–40 km of angling reserve area, where one can fish. Some other streams of Himachal include Sutlej, Ravi, Tirthan, Sainj, Lambadug, Giri, Rana, Nugal Gai, Baner, Bata, etc. The major fishes available in these streams are Trout, Mahseer, *Nemacheilus* spp., *Barilus* sp, *Schizothoracids*, *Crossocheilus* sp. *Glyptothorax* spp. etc. Fishing in these streams is regularized under the state fisheries Act. In trout water, licenses only for rod and line are permitted, while in general water both rod and line as well as cast netting is allowed. Angling rules are liberal with a nominal fee. Angler is permitted to catch six trouts a day on each license; however a trout should not be less than 40 cm in size. Trout fishing close season lasts from 1st November to last day of February each year. Some of the fishing sites famous for trout angling are Rohru, Chirgaon, Seema, Mandil, Saandhsu, Tikri, Dhamwari, Baspa, Sangla, Kupa, Badseri and Barot located at Luhandi, Puran hatchery, Lachkkandi, Tikkar, Balh, and Kamand. Besides Barot, the entire reservoir from Pandoh Dam to Aut on the Mandi-Manali national highway is also considered good for trout fishing. Brown Trout can be found in the streams that feed the Beas River in the Kullu Manali region. The best beats may be found in Larji Valley, which is located at the confluence of the rivers Larji and Tirthan. On the other side, the trout stream Sainj flows (Mandal et al., 2018).

4 Potential of Sport Fishery in Uttarakhand

The hill state of Uttarakhand in the western Indian Himalayan Region has a size of 53,483 km^2 and is made up of around 86% hills and 14% plains. Uttarakhand is divided into two administrative divisions, Kumaun and Garhwal, with a total of 13 districts. The state's distinct environment and terrain provides substantial natural aquatic resources in the form of 2686 km of rivers and streams, 132 highland lakes (encompassing 297 ha), and 14,120 ha of man-made reservoirs (Department of Fisheries, Govt. of Uttarakhand, 2021; Verma, 2013).The state of Uttarakhand continues to offer abundant options for adventure sports and for the recreational activities like fishing and angling. Angling/sport fishing is made available on some of the most prominent and pristine lakes and rivers of the State. The ideal time for fishing is from September to May as the temperature remains quite pleasant and mild. Fishing and angling is extensively enjoyed in the resorts settled in Upper Ramganga River in Corbett National Park. It is increasingly becoming a popular recreation sport for tourists in Uttarakhand. Several resorts and hotels in Nainital have introduced angling as one of the activities for their guests to indulge in. For carrying out angling activities, angling permits for national and international tourists are a must. The beats are identified under Mahseer and Trout Zones. Angling spots in Garhwal

Himalayas include the Bhagirathi and Alaknanda River, Tons, Song, Nayar, Aglar, Amlawa, Yamuna, Dodital. Angling spots in Kumaon Himalayas include Ramganga, Sattal, Corbett National Park, Ramganga River, Kosi River, Sonanadi, and Sharda River. The golden mahseer, also known as the Himalayan mahseer (*Tor putitora*), is the most popular game fish among anglers in India due to its superior fishing qualities. Because angling was a popular British pastime, Sir H. Ramsay filled the lakes of the Kumaon hills with fish in 1858 for sport fishing (Walker, 1888). The state has a total combined length of 10,928 km of river network, of which 3151 km has the potential to sustain acceptable habitat for brown trout based on elevation and existing thermal conditions (Rawat et al., 2011).

Nandakini, Pranmati Gad, Dhauli Ganga, Laisar Gad, Bhilangna River, Har Ki Dun, Pinder (Singh et al., 1983), Asiganga, Lake Dodi Tal, Balkhila Gad, and Madhu Ganga are some of the important stretches of river and their upstream areas harboring the brown trout (*Salmo trutta fario*). In the northern section of Uttarakhand, the Nandakini, Dhauli Ganga West, Laisar Gad, Bhagirathi, Bhilangna, Asiganga, Pranmati Gad, Bal Ganga, Badiyar Gad, Gomati, Balasuti, and Pinder rivers can provide suitable habitat for brown trout (Nayak et al., 2014). Besides, Sattal, Khurpatal, and Bhimtal in Nainital District, Kali River in Pithoragarh, Pancheshwar on the bank of Mahakali River in Kumaon are among the few places to enjoy this activity in Uttarakhand. The beautiful Dodital in Garhwal Region, which is one of the popular trekking destinations, also allows the opportunity for angling and fishing. It should be noted that fishing in Uttarakhand is only allowed on "Catch and Release" basis, and permission has to be taken in advance from the concerned authorities.

5 Potential of Sport Fishery in Northeast Region of India

Northeast India is rich in natural resources, including substantial freshwater supplies that sustain diverse aquatic life, including cold-water fish species. The region has a total water surface area of around 13,000 km^2 and comprises numerous important rivers, including the Brahmaputra, Barak, and Subansiri, as well as their tributaries, which are home to a variety of cold-water fish species. The Himalayas and the Indo-Burma region are two of the 34 biodiversity hotspots identified by Conservation International. The North East Region shares its fish fauna predominantly with that of the Indo-Gangetic fauna and to a small extent with the Burmese and South China fish fauna. The Manas river in Assam affords excellent wild life viewing in addition to Mahseer fishing. Manas river in Manas National Sanctuary also has large snow trout, which can be a potential spot for angling. The Jia Bharali in Assam is considered an important place for angling. Arunachal Pradesh has 340 km of trout streams. Besides Meghalaya also has huge potential for recreational fishery, where angling is a common scene in almost all the rivers and reservoirs of the state.

6 Suggestions to Promote Sport Fishery in the Himalayan Region

1. The conservation plan/protocol for mahseer, snow-trout may be framed.
2. Mapping of Indian waters having stock of mahseer and snow-trout and bring them on the international map of angling-tourism.
3. Approach road to the fishing sites, rest camps, angler's hut should be developed.
4. Aritificial propagation of mahseer and snow-trouts should be taken up for large scale production.
5. Fisheries acts and regulations should be adopted strictly to augment fishery of endangered fish species like mahseer and snow-trouts.
6. Angling competitions may be organized annually with publicity by news media, pamphlets, audio and video visions, etc., to attract maximum numbers of tourist/anglers to visit the area.

Thus, if the potential of sport fishery in the Himalayas of India is utilized judiciously, the economy of this region can be uplifted. It can also help to create employment opportunities for the rural unemployed educated youth.

7 Conclusion

In Indian Himalayan region, the growth of sport fishery has great potential to stimulate local economies and generate job possibilities, especially for young people from rural areas. The area has a lot of potential to draw anglers and tourists because of its varied and rich cold-water ecosystems. A deliberate approach comprising the preservation of important fish species, the construction of infrastructure, artificial propagation, and the stringent enforcement of regulations is necessary to fully achieve this potential. Tourists and professional fisherman can be drawn in even more by boosting publicity and organizing fishing competitions. In the Himalayan region, sport fishing has the potential to develop into a profitable and sustainable industry with careful management.

References

Canyurt, M. A., & Akhan, S. (2009). *Development and situation of trout culture in Turkey* (pp. 90–94). Agricultural Sciences (Crop Sciences, Animal Sciences).

Department of Fisheries, Govt. of Uttarakhand. (2021). https://fisheries.uk.gov.in.

Jhingran, V. G., & Sehgal, K. L. (1978). *Coldwater fisheries of India* (p. 238). Inland Fisheries Society of India.

Mandal, A., Das, S. K., Biswas, B., & Kairnar, S. O. (2018). Future scope of Sport Fisheries Development in India: A review. *Indian Journal of Animal Health, 57*(1), 1–16.

Nayak, A. K., Pant, D., Kumar, P., Mahanta, P. C., & Pandey, N. N. (2014). GIS-based aquaculture site suitability study using multi criteria evaluation approach. *Indian Journal Of Fisheries, 61*(1), 108–112.

Rawat, M. S., Bantwan, B., Singh, D., & Gusain, O. P. (2011). Status of brown trout (*Salmo trutta fario* L.) in Garhwal Himalaya with a note on its morphometric characteristics. *Environment Conservation Journal, 12*(3), 47–52.

Sehgal, K. L. (1974). Trout culture in India, its present status and future scope. *Journal of the Inland Fisheries Society of India, 6*, 185–193.

Sehgal, K. L. (1999). Coldwater fish and fisheries in the Western Ghats, India. In *Fish and fisheries at higher altitudes: Asia*. FAO.

Singh, H. R., Joshi, C. B., & Prasad, R. (1983). The trout fishery of Garhwal: Present status, management problems and prospects. In T. V. Singh & J. Kour (Eds.), *Studies in eco-development: Himalayas, mountain and men*. Print House.

Verma, R. (2013). Water and fishery resources of Kumaun Central Himalaya, India. *Discovery, 8*(19), 7–12.

Walker, W. (1888). Angling in Kumaun Lakes, Calcutta.

Sea Game Fishing in Western Ghats

7

Abstract

The Western Ghats of India, a UNESCO World Heritage Site, offer exceptional sport fishing opportunities for anglers seeking a diverse array of game fish species. This region, stretching along the southwestern coast of the country, is home to numerous rivers, streams, and reservoirs that harbor prized fish like the Mahseer, Barramundi, and various species of Trout. Key destinations in the Western Ghats include the Cauvery River in Karnataka, the streams of Munnar in Kerala, and the Kali River along the Karnataka-Goa border. These waters provide a thrilling angling experience, with the best seasons varying by location. However, the long-term sustainability of these game fish populations is under threat from habitat degradation, pollution, and overfishing. Effective conservation measures, such as catch-and-release practices and the establishment of protected areas, are crucial to ensure the continued enjoyment of this cherished sport fishing tradition in the Western Ghats.

Keywords

Western ghats · Cauvery river · Munnar · Conservation · Sustainability

1 Introduction

The Western Ghats is a mountain range in India that stretches over six states along the western coast. The region is known for its rich biodiversity and is home to many species of flora and fauna, including several endemic species. In addition to its natural beauty, the Western Ghats is also known for its fishing opportunities, including sea game fishing. The region is home to many species of game fish, including barracuda, tuna, kingfish, grouper, and snapper. The Arabian Sea, which lies to the west of the Western Ghats, provides an ideal habitat for these fish, and the region is

© The Author(s), under exclusive license to Springer Nature Singapore Pte Ltd. 2024
S. T. Mushtaq et al., *Angling in India*,
https://doi.org/10.1007/978-981-97-8876-7_7

well-known for its sport fishing opportunities. Some popular locations for sea game fishing in the Western Ghats include Goa, Mangalore, and Kochi. These areas offer a variety of fishing options, from deep sea fishing to shore fishing. Many tour operators and fishing charters offer guided trips to these locations, providing all necessary equipment and expertise to ensure a successful fishing experience. It is important to note that certain species of fish in the Western Ghats are protected, and fishing regulations may vary depending on the location. It is important to check with local authorities and fishing charters before embarking on a sea game fishing trip in the Western Ghats.

Additionally, it is important to practice responsible fishing practices, such as catch and release, to help preserve the region's marine ecosystem for future generations. The Western Ghats ranges in western India run for about 1600 km and have an average altitude of 1200 m (max. 2339 m) with 102 species of fish. The three major rivers draining west are Godavari, Krishna, and Cauvery, the last river being famous for its sport and recreational fishing for mahseer (*Tor khudree*). The fish stocks have been under an intensive fishing pressure from subsistence and sport/recreational fishermen, and this has led to their decline. In certain sections of the Cauvery, under the management of fishing clubs and associations, the stocks are now successfully managed to support the ever increasing numbers of recreational fishermen. Such sections are regularly stocked, and fishing regulations are strictly enforced. The upper streams of the Western Ghats in Karnataka, Tamil Nadu, and Kerala, support stocks of rainbow trout. Both commercial and recreational fisheries are practised in the Western Ghats. Commercial catches are taken from the main rivers when they meander through plains, and from some lakes and reservoirs.

The past information on Mahseer fishing in the Western Ghats is primarily based on the result of a Trans World Fishing Expedition (TWFE), which took place in 1978, and on Boote Mission which took place the same year. The Expedition was arranged by the Indian Tourism Development Corporation in collaboration with Air India. The team visited a number of regions in the country in search of the big Mahseer. It collaborated closely with the Wildlife Association of South India (WASI) at Bangalore. It was at the time of a sharp decline in catches of large Mahseer that the National Commission on Agriculture recommended a comprehensive survey of rivers and streams known to have Mahseer stocks. In 1976, WASI leased about 22 km of the Cauvery, 60–80 km distant from Bangalore, for conservation of *Tor khudree.*

The Coorg Wild Life Society, Madikere, is another voluntary organization which is engaged in protecting mahseer fishery in the Cauvery, with a lease on 28 km of this river. This Society has been stocking young mahseer in this stretch since 1993. It also organizes sport fishing and maintains fish catch statistics. The anglers are governed by similar rules and regulations as followed by WASI. There are several types of game fishing opportunities available in the Western Ghats, each with its own unique challenges and rewards. These are as follows:

1.1 Deep Sea Fishing

This type of fishing involves venturing out into the open sea, where anglers can target larger species such as tuna, marlin, and sailfish. Deep sea fishing requires specialized equipment and knowledge, and is often done with the help of experienced fishing guides.

1.2 Shore Fishing

This type of fishing can be done from the shore or nearshore areas, such as estuaries or jetties. Shore fishing is accessible to a wider range of anglers and can be a great way to target smaller species such as barracuda, snapper, and groupers.

1.3 Kayak Fishing

Kayak fishing is a popular option for anglers who want to get up close and personal with the fish. Kayaks allow anglers to navigate shallow waters and tight spaces, making it easier to reach areas that are inaccessible by larger boats (Fig. 7.1).

Fig. 7.1 Kayak fishing. Floodmfx, CC BY-SA 4.0. https://creativecommons.org/licenses/by-sa/4.0. Via Wikimedia Commons

1.4 Fly Fishing

Fly fishing is a more challenging and specialized form of fishing that requires a specific set of skills and equipment. The Western Ghats offer opportunities for fly fishing for species such as mahseer.

1.5 Night Fishing

Night fishing can be a unique and exciting experience for anglers looking to target species such as snapper and barracuda, which are more active at night. It can be done from shore or on a boat, and requires specialized equipment such as underwater lights.

2 Fish Species for Sea Game Fishing in Western Ghats

The Western Ghats region in India has a variety of fish species that are suitable for sea game fishing. Some of the popular fish species for sea game fishing in the Western Ghats include Barracuda, Mahi-Mahi, Tuna, Snapper, Grouper, Kingfish, Sailfish, Marlin, Wahoo, Trevally, Exotic Rainbow Trout, and Mahseer. It is important to note that fishing regulations and guidelines may vary in different areas and it's always important to follow the guidelines set by local authorities. Additionally, catch and release practices should be followed to preserve the fish population in the area. Diversity and distribution of exotic rainbow trout: It is present in streams of the Bababudan Range in Karnataka, in Nilgiris and Palani Hills in Tamil Nadu, and in High Range in Kerala. It was perhaps the Nilgiris Game Association which took the initiative of stocking the streams of the Blue Mountains Diversity and distribution of mahseer in Western ghats Deccan mahseer in Cauvery river.

The Deccan Mahseer (*Tor khudree*) is known in vernacular Kannada as Bilimeenu (white fish) and since long is associated with heritage and belief. It is considered as Devara Meenu (Divine fish) in some parts of Karnataka and is fed by local people in the sanctuaries and rivers associated with the temples. The Mahseer is distributed in the Siwalik Himalaya and lower reaches of the uplands of the Deccan Plateau. Although this fish occurs at several reaches in the upper stretch of river Cauvery, three reaches are important from the abundance point of view, viz., Kushalnagar (Walnoor to Kushalnagar), Ramanathapur, and Bheemeswari (Shivasamudram falls to Mekedatu).

3 Rules and Regulations for Sea Game Fishing in Western Ghats

Sea game fishing in the Western Ghats region of India is governed by a set of rules and regulations to protect the marine ecosystem and ensure sustainable fishing practices. Some general rules and regulations include:

3.1 Fishing Permits

Before engaging in sea game fishing, you need to obtain a fishing permit from the local authorities.

3.2 Protected Species

Certain species of fish, such as certain types of sharks, are protected by law and cannot be caught or kept. You should familiarize yourself with the protected species in the area and avoid catching them.

3.3 Catch Limits

There are catch limits on certain species of fish to prevent overfishing. One should be aware of these limits and adhere to them.

3.4 Gear Restrictions

Some areas may have restrictions on the type of fishing gear that can be used, such as size limits for hooks and nets.

3.5 Size Limits

Some species of fish have size limits, which means that fish below a certain size cannot be kept. One should familiarize oneself with the size limits for the species being targeted.

3.6 Catch and Release

Whenever possible, catch and release practices should be followed to conserve the fish population. If you do catch a fish that you plan to release, handle it carefully to minimize injury and maximize its chances of survival.

3.7 Environmental Protection

When fishing in the Western Ghats, one should also take care to protect the marine environment. This means avoiding littering and using biodegradable fishing gear whenever possible. It's important to note that specific rules and regulations may vary depending on the area one is fishing in. It's always best to check with local authorities for the latest regulations and guidelines before engaging in sea game fishing in the Western Ghats.

4 Conclusion

The Western Ghats, renowned for its rich biodiversity and stunning landscapes, offers excellent sea game fishing opportunities. This region, with its diverse aquatic habitats, supports a wide variety of game fish, including barracuda, tuna, kingfish, grouper, and snapper. Popular fishing locations like Goa, Mangalore, and Kochi provide various fishing options, from deep-sea fishing to shore fishing, with many tour operators offering guided trips.

The region's fishing resources, including the Arabian Sea and numerous rivers and lakes, face significant pressure from both subsistence and recreational fishing. Efforts by organizations like the Wildlife Association of South India and the Coorg Wild Life Society have led to successful management and conservation practices, particularly in the Cauvery River. The introduction of regulations and restocking programs has helped sustain fish populations, benefiting both the environment and the angling community. For sustainable sport fishing, it is crucial to adhere to local regulations, practice catch and release, and follow responsible fishing practices. By doing so, the unique marine ecosystem of the Western Ghats can be preserved for future generations, supporting both biodiversity and local economies through tourism and recreational activities.

Sea Game Fishing in the Andaman and Nicobar Islands

8

Abstract

The extensive and varied marine life of the Andaman and Nicobar Islands draws anglers seeking excellent sea game fishing chances. Important spots like Long Island, Neil Island, Havelock Island, Richie's Archipelago, and others offer top-notch fishing for marlin, sailfish, tuna, barracuda, wahoo, and other species. With their gear and knowledge, fishing charters from well-known locations like Port Blair and Havelock Island improve the fishing experience. Fishing is best done by trolling, casting, or jigging; the best months to fish are November through April. Sustainable fishing and the preservation of the maritime environment are ensured by adherence to licenses, laws, and conservation techniques.

Keywords

Andaman and Nicobar Islands · Game fishing · Fishing charters · Fish species · Sustainable fishing practices

1 Introduction

The Andaman and Nicobar Islands are home to a wide variety of fish species, including marlin, sailfish, tuna, barracuda, wahoo, and many more. The islands offer excellent opportunities for anglers to indulge in both deep-sea and sport fishing. The tropical islands located off the coast of Neil Island, Havelock Island, Richie's archipelago, and Long Island provide the apt fishing grounds with great catches and never cease to amaze enthusiasts. These islands are a unique destination for game fishing, where one can find so many fishing spots such as Ruth Land, Cinque Island, and the islands that surround Havelock (Fig. 8.1). Big boats and small boats are available for game fishing.

© The Author(s), under exclusive license to Springer Nature Singapore Pte Ltd. 2024
S. T. Mushtaq et al., *Angling in India*,
https://doi.org/10.1007/978-981-97-8876-7_8

Fig. 8.1 Havelock island, Andaman and Nicobar. © Vyacheslav Argenberg CC BY 4.0. https://creativecommons.org/licenses/by/4.0. Via Wikimedia Commons

2 Fishing Charters

To go game fishing in the Andaman and Nicobar Islands, it is advisable to book a fishing charter. Several charter operators are available in Port Blair, Havelock Island, and other popular tourist destinations. These charters provide boats, equipment, and experienced guides to enhance your fishing experience.

3 Fish Species

The Andaman and Nicobar Islands offer a diverse range of fish species that can be caught while game fishing. Here are some notable species that anglers often target:

3.1 Marlin

The Andaman and Nicobar Islands are known for their excellent marlin fishing opportunities. The region is home to both blue marlin and black marlin, which are highly sought-after game fish due to their size, strength, and acrobatic fights.

3.2 Sailfish

Sailfish are another popular species targeted by game fishermen in the Andaman and Nicobar Islands. Known for their incredible speed and agility, sailfish provide thrilling battles when hooked. They are often caught using trolling techniques with live bait or artificial lures.

3.3 Tuna

Several tuna species inhabit the waters around the islands, including yellowfin tuna and dogtooth tuna. Tuna are prized for their strength and speed, making them challenging adversaries for anglers. Trolling and casting with poppers or metal jigs are common techniques used to catch tuna.

3.4 Barracuda

Barracuda are aggressive predators and can be found in abundance around the islands. They are known for their sharp teeth and fast strikes. Barracuda are often caught using a variety of techniques, including trolling, casting, and jigging.

3.5 Wahoo

Wahoo are prized game fish known for their speed and powerful runs. They are often found near drop-offs, reefs, and underwater structures. Trolling with lures at high speeds is an effective method for targeting wahoo.

3.6 Mahi-Mahi (Dorado)

Mahi-Mahi, also known as dorado or dolphin fish, are vibrant and acrobatic game fish. They are frequently encountered in the waters of the Andaman and Nicobar Islands. Mahi-Mahi are known for their dazzling colors, and they provide exciting fights when hooked. They are commonly caught using trolling techniques with lures or live bait.

3.7 Giant Trevally

Giant Trevallies are powerful and aggressive fish that can be found in the waters around the islands. They are known for their strength and explosive strikes. Casting poppers, jigs, or live bait is a popular method to entice these fishes.

4 Fishing Techniques

The fishing techniques used in the Andaman and Nicobar Islands depend on the target species. Trolling, both with artificial lures and live bait, is commonly employed for species like marlin and sailfish. Casting poppers and jigs are also effective for catching pelagic fish.

5 Fishing Seasons

The best time for game fishing in the Andaman and Nicobar Islands is generally from November to April when the seas are calm and the weather is pleasant. However, certain fish species may have specific peak seasons, so it is recommended to check with local fishing operators for the most up-to-date information.

6 Permits and Regulations

It is important to obtain the necessary permits and follow fishing regulations while engaging in game fishing in the Andaman and Nicobar Islands. Fishing permits can be obtained from the Forest Department or local authorities. It is essential to adhere to catch limits, size restrictions, and any other regulations to ensure sustainable fishing practices.

6.1 Catch Limits and Size Restrictions

The government sets catch limits and size restrictions on certain fish species to ensure sustainable fishing practices and protect vulnerable populations. It is crucial for anglers to be aware of and adhere to these regulations. Catch limits may vary depending on the species and can be subject to change, so it is advisable to stay informed about the latest guidelines.

6.2 Protected Species

Certain species are protected under Indian wildlife protection laws, and it is strictly prohibited to target or catch them. These protected species may include marine mammals, sea turtles, and specific fish species. It is essential to familiarize yourself with the list of protected species and ensure they are promptly released if accidentally caught.

6.3 Catch and Release

Responsible catch and release practices are encouraged to preserve fish populations and the marine ecosystem. Anglers are advised to release undersized fish, endangered or protected species, and any catch that is not intended for consumption. Proper handling techniques and the use of barbless hooks can minimize harm to the fish during the release process.

6.4 Gear Restrictions

The use of certain fishing gear, such as gill nets or other indiscriminate fishing methods, may be prohibited to protect marine resources and prevent bycatch. It is important to comply with gear restrictions specified by the local authorities and use fishing techniques that minimize environmental impact.

6.5 Conservation Efforts

Responsible and sustainable fishing practices are encouraged to protect the marine ecosystem in the Andaman and Nicobar Islands. Anglers are advised to practice catch and release whenever possible, particularly for threatened or protected species. By doing so, they can contribute to the long-term preservation of the region's fish population.

7 Conclusion

Sea game fishing in the Andaman and Nicobar Islands offers an exceptional experience due to the region's rich marine biodiversity and favorable fishing conditions. With a variety of target species and effective fishing techniques, anglers can enjoy thrilling and successful fishing trips. It is crucial to follow local regulations, obtain necessary permits, and practice responsible fishing to preserve the marine environment. Through sustainable practices and adherence to conservation efforts, the Andaman and Nicobar Islands can continue to be a top destination for sea game fishing enthusiasts.

Sea Game Fishing of Lakshadweep Islands

9

Abstract

Lakshadweep, an archipelago in the Arabian Sea, is renowned for its rich marine biodiversity and as a popular destination for sea game fishing. Anglers can target species such as tuna, marlin, sailfish, barracuda, and sharks through various fishing techniques including bottom fishing, trolling, popper fishing, jigging, and fly fishing. The best fishing season spans from October to May. Strict regulations, including permits, catch limits, and protected areas, ensure sustainable fishing practices. Engaging local guides enhances safety and success. Lakshadweep's commitment to conservation ensures the preservation of its unique marine ecosystem for future generations.

Keywords

Lakshadweep · Sea game fishing · Fishing techniques · Marine conservation · Fishing regulations

1 Introduction

Lakshadweep, an archipelago located in the Arabian Sea (Fig. 9.1), is known for its rich marine biodiversity and also happens to be a popular destination for sea game fishing. There are several species of fish that can be found in the waters surrounding Lakshadweep which include tuna, marlin, sailfish, barracuda, and shark. Marine game fishing in Lakshadweep is either done through organized fishing trips or by hiring local fishing boats. As some species of fish are protected in the area, it is essential to follow local regulations and guidelines to preserve the marine ecosystem. It is also advisable to engage the services of experienced and licensed guides or operators to ensure safety and maximize the chances of a successful fishing trip.

© The Author(s), under exclusive license to Springer Nature Singapore Pte Ltd. 2024

S. T. Mushtaq et al., *Angling in India*,
https://doi.org/10.1007/978-981-97-8876-7_9

Fig. 9.1 Lakshadweep archipelago. Jeff Schmaltz, MODIS Land Rapid Response Team, NASA GSFC, Public domain, via Wikimedia Commons

2 Types of Game Fishing in Lakshadweep

There are several types of game fishing that can be enjoyed in Lakshadweep which include:

Fig. 9.2 Bass popper. Curtis Fry, CC BY 2.0. https://creativecommons.org/licenses/by/2.0. Via Wikimedia Commons

2.1 Bottom Fishing

It is a popular fishing method where the bait is placed on the ocean floor and anglers wait for fish to take the bait. Fish species such as groupers, snappers, and emperor fishes are caught using this method.

2.2 Trolling

Trolling, which involves dragging a baited line behind a moving boat to attract predatory fish is the most common form of game fishing in Lakshadweep. Fishes such as tuna, sailfish, and marlin are caught using this method.

2.3 Popper Fishing

Popper fishing involves casting a popper lure into the water and then retrieving it with a jerking motion to imitate a wounded baitfish (Fig. 9.2). This technique is popular for catching species such as Giant Trevally and Dogtooth Tuna.

2.4 Jigging

Jigging involves using a heavy metal lure to create a vertical motion in the water and is effective for catching species such as Amberjack and Grouper.

2.5 Fly Fishing

Fly fishing involves using a fly lure that imitates the appearance and movement of small baitfish or insects. This technique can be used to catch a variety of species such as Bonefish and Barracuda.

The best time for game fishing in Lakshadweep is between October and May, during the post-monsoon and winter seasons. During this time, the weather is generally mild, and the seas are calm and ideal for fishing. The water temperature is also suitable for several species of fish, including tuna, sailfish, and marlin. From June to September, the monsoon season brings heavy rains and strong winds, making fishing challenging and dangerous as a result of which most fishing operators and guides in Lakshadweep suspend their services. Therefore, it's advisable to check with local operators or fishing guides for the best times to fish based on current conditions. Besides, timing of the fishing season can vary depending on weather patterns, and fish migration patterns can also affect the availability of different species.

3 Game Fishing Species in Lakshadweep

Lakshadweep is known for its diverse marine ecosystem and some of the popular game fishing species that can be found in the waters include:

3.1 Giant Trevally (*Caranx ignobilis*)

Giant Trevally are a popular game fish in Lakshadweep known for their strength and fighting ability and can be caught using popping or jigging techniques. This powerful fish is a popular target for sport fishermen in Lakshadweep.

3.2 Tuna (*Thunnus* spp.)

Several species of tuna can be found in the waters around Lakshadweep, including Yellowfin Tuna, Big eye Tuna, and Skipjack Tuna. These are prized for their speed and strength and are popular for their firm and flavorful meat. These can be caught using trolling or jigging techniques.

3.3 Sailfish (*Istiophorus platypterus*)

Sailfish are known for their incredible speed and acrobatic displays when hooked and are typically caught using trolling or live bait fishing techniques.

3.4 Barracuda (*Sphyraena* spp.)

Barracudas, a group of predatory fish known for their sharp teeth and aggressive strikes and are common game fishes in the waters around Lakshadweep. They can be caught using trolling or casting techniques and are popular for their firm, white flesh.

3.5 Wahoo (*Acanthocybium solandri*)

Wahoo are fast-swimming game fish that are prized for their firm, white flesh and are often caught by trolling with lures or live bait.

3.6 Marlin (*Makaira* spp.)

Several species of marlin (Blue Marlin and Black Marlin) can be found in the waters around Lakshadweep. These powerful fish are known for their long, hard runs and acrobatic jumps. These are a prized catch for many game fishermen and are typically caught using trolling or live bait fishing techniques.

3.7 Dogtooth Tuna (*Gymnosarda unicolor*)

Dogtooth tuna/doggies, are a group of large predatory fish that can be caught using popping or jigging techniques. These fish are prized for their strong fighting ability and delicious meat.

4 Regulations Governing Recreational Fishing

Recreational fishing in Lakshadweep is regulated by the Lakshadweep Administration, and visitors are required to obtain a permit before engaging in any fishing activities. The Lakshadweep administration, which is the governing body of the islands enforces strict catch limits and encourages anglers to practice catch-and-release fishing to conserve the marine ecosystem. This means that all fish caught must be released back into the water immediately. Fishing is only allowed in designated areas and during specific times of the year.

In order to protect the unique diversity of these islands, the state administration has also identified certain areas as Marine Protected Areas (MPAs), where fishing is prohibited. The use of certain types of fishing gear, such as gill nets, is also prohibited in the Islands. Fishing permits are required for all sport fishermen, and the number of permits issued is limited to ensure sustainable use of fishery resources. There are several fishing tours and charters available in Lakshadweep that cater to both novice and experienced anglers. These tours provide all the necessary equipment and gear, including fishing rods, reels, bait, and lures. Additionally, most tours offer experienced guides who can help one find the best fishing spots and offer tips on how to catch different types of fish. Many fishing tours also offer the chance to explore the surrounding islands and enjoy activities such as snorkeling, scuba diving, and island hopping. Some popular spots for sea game fishing in Lakshadweep include Agatti Island, Kadmat Island, and Kavaratti Island. Each of these islands offers a unique fishing experience, with different types of fish and fishing techniques. All sport fishermen are expected to follow the regulations and guidelines issued by the local authorities and cooperate with them to ensure the protection of the marine ecosystem.

5 Rules and Plans

5.1 The Lakshadweep Marine Fishing Regulation (LMFR, 2000) Act

This law regulates all fishing activities in the Lakshadweep Islands and defines the rights and obligations of fishermen, fishing boat owners, and fishing cooperatives. The law also establishes penalties for violations of the regulations.

5.2 The Lakshadweep Marine Conservation Rules

These rules define the criteria for the identification and management of Marine Protected Areas (MPAs) in the Lakshadweep Islands. The rules also regulate the use of fishing gear and the collection of marine resources in these MPAs.

5.3 The Lakshadweep Fisheries Management Plan

This plan outlines the strategies and actions for the management and conservation of fishery resources in the Lakshadweep Islands. It includes measures such as the promotion of sustainable fishing practices, the establishment of closed fishing seasons and areas, and the monitoring and surveillance of fishing activities.

5.4 The Lakshadweep Coral Reef Conservation Project

This project aims to conserve and manage the coral reefs and associated marine biodiversity in the Lakshadweep Islands. It includes activities such as the establishment of marine protected areas, the restoration of damaged coral reefs, and the development of sustainable livelihood options for local communities.

Lakshadweep offers a unique destination for marine game fishing enthusiasts, with its rich biodiversity and various fishing techniques suitable for all skill levels.

The fishing regulations in Lakshadweep reflect the government's commitment to sustainable development and the protection of the unique marine ecosystem in the Lakshadweep Islands. These regulations and initiatives are strictly enforced, and certain fish species may be protected or restricted. Engaging in responsible fishing practices such as Catch and Release and respecting protected areas helps maintain the delicate balance of the ecosystem so that future generations will continue to enjoy the underwater world and the thrill of game fishing in this area.

6 Conclusion

Sea game fishing in Lakshadweep offers a diverse and exciting experience for anglers, with numerous species and fishing techniques to explore. The region's rich marine biodiversity and well-regulated fishing practices make it a premier destination for fishing enthusiasts. By adhering to local regulations, obtaining necessary permits, and practicing responsible fishing, anglers can contribute to the sustainable management of Lakshadweep's marine resources. The involvement of experienced guides and operators further enhances the fishing experience, ensuring both safety and success. Through ongoing conservation efforts, Lakshadweep aims to preserve its marine ecosystem, allowing future generations to enjoy the thrills of sea game fishing in this unique archipelago.

Casting 10

Abstract

Casting is an essential skill in fishing, requiring knowledge of various techniques and gear to effectively attract and catch different fish species. Mastery of different casting methods, from basic overhead and sidearm casts to specialized techniques like double haul and tuck casts, allows anglers to adapt to diverse fishing conditions and target specific fish. Understanding the nuances of cast and retrieve techniques further enhances the angler's ability to mimic natural prey movements and entice predatory fish. Safety precautions are crucial to prevent injuries, emphasizing the importance of awareness, and careful handling of equipment. Overall, proficiency in casting techniques is fundamental for successful and safe fishing.

Keywords
Casting · Sidearm cast · Overhead cast · Handling · Fishing

1 Introduction

Casting is a method of fishing in which a bait or lure is thrown in water using a flexible fishing rod and the line is retrieved to play the lure in order to create an action of the lure to attract the predatory fish to prey (Fig. 10.1). Casting technique is mostly applied to ultralight fishing by using various kinds of artificial lures. Casting is often used for fishing near beaches, rivers, lakes, and swamps. There is also a casting technique that is applied in the sea but near shallow corals. The fish targeted by casting techniques are small fish such as Giant Snakehead, Mahseer, Bluefin Trevally, Diamond Trevally, Barramundi, Snakehead, Tarpon, Mangrove Jack, and many others. The lure or artificial bait used varies depending on the target of the fish being caught. Examples include minnow, swimbait, crankbait, spoon, spinner,

© The Author(s), under exclusive license to Springer Nature Singapore Pte Ltd. 2024
S. T. Mushtaq et al., *Angling in India*,
https://doi.org/10.1007/978-981-97-8876-7_10

Fig. 10.1 Casting—an angler in action (Picture credit: Ubaid Wani)

frogs, and also soft lure bait such as worms and so on. In casting technique, the anglers can also use the cast jig. The action that is produced by each lure above varies depending on the type of lure used. Globally the sport is supervised by the International Casting Sport Federation (ICSF) which was founded in 1955 and as of April 2014 has member associations in 31 countries.

2 Casting Technique

To properly grip a fly rod, it is held securely with the thumb extending along the spine of the rod, opposite of the reel. One should not hold it too tightly, as by doing so one might overpower the casts. The thumb will impart the power of the cast, so placing it along the spine of the rod will keep the power where it needs to be. The thumb points where the rod tip goes is where the line will shoot. With the other hand, hold the fly line. The non-dominant hand controls the amount of line that is being cast. The other control point is under the pointer finger of the dominant hand. This serves as a pinch point for the line against the rod. First, one needs to bend the rod or load the rod to shoot the line with help of the weight of the fly line. Therefore, one needs to have the fly line out of the guides first to load the rod.

The other key point is that when casting, one will be making a loop of the fly line to achieve the best aerodynamics and hence distance and accuracy. As one moves the fly rod, the energy of the cast is transferred to the fly line, and the loop of the line will unfurl and delicately present the fly on the water. The line must unfurl for best results with both the back cast and the forward cast. For this reason, when starting out, one should position oneself in such a way that one can watch their back cast as well.

Another essential point is that to create the tightest loops and the most efficient energy transfer, the rod tip must move straight, not an arc. To do this, the tip should move from roughly 10 o'clock to 2 o'clock. These are the standard positions for the basic overhead cast. Because the physics of casting is energy transfer, the casting stroke should be an acceleration followed by a crisp stop. This stop helps load the rod and transfer energy to the line. To make this movement, both the forearm and wrist are used. The traditional way is to use your forearm and lock the wrist, just like a hammer to drive a nail. The new approach is all about flicking the wrist. Many people prefer a wrist snap at the end of the motion for a smooth acceleration and quick stop. The main point is keeping a smooth acceleration followed by a crisp stop.

3 Getting into the Cast

For the overhead casting stroke, start one step at a time. Stand sideways and watch it in both directions. It is just a three-count movement. On count one, snap the line back. Watch it unfurl. Count two, bring it forward and watch it unfurl. Remember to pause in between the back and forward cast. If you hear a snap, you are not pausing, and you will snap off flies when fishing. Once you are comfortable with moving from count one to count two and back and forth, it is time for count three. As you bring your rod forward, for count three, drop the rod tip to 9 o'clock. This will lay the unfurled line down on the water. With practice, one should be able to cast up to 30 feet consistently. This distance is sufficient to catch most fish.

4 Types of Gear Used

Freshwater anglers typically use lightweight, faster-action rods for popular mid-sized game fishes such as black bass or trout, while sturdier, heavier rods are used for larger, stronger fish. When casting light rods, sidearm casting techniques are typically used. If more distance is needed, an overhead cast is more common. Saltwater anglers usually use heavier rods and lines, as they often use baits and lures that are bigger and heavier than those used on freshwater. For surfcasting and specialized two-handed casting techniques, heavier rods are used to provide the added distances required for the baits/lures to get past the surfs and reach inshore fish feeding near the upwellings. In these casts, the angler utilizes the entire body, rather than just the arms, to deliver the momentum of the cast, which may travel many hundreds of meters/feet.

5 Types of Casts

5.1 Overhead Cast

Overhead casting technique is commonly used by many casting anglers. The overhead cast is the basic fly fishing cast. It has evolved over the years primarily due to advances in equipment. In the "olden days" before graphite rods, a lot more wrist movement was involved in the cast. Now, there is more arm and shoulder movement in order to "load the rod" to better utilize equipment (graphite). However, the basic concept is still the same. Bring the fly line overhead and behind you, then cast it front of you to the desired target. The ideal cast should have the line tight at all times and the line should form tight loops at the end of the back and forward casts. The line should remain horizontal to the ground at all times. Overhead casting involves the following steps and some careful considerations as well. The rod handle should be gripped with the thumb on top, pointing straight down the rod. The rod tip should be pointing slightly out to the side (this keeps the fly at a safe distance from the caster) and low to the water to start the cast. Lift the rod with a slow gliding lift to about the 10.30 position then increase the speed back to 12.30 (just past vertical) and stop (at the stop, the thumb should be no further back than the ear and should be below it). Pause for the line to extend behind. Cast the rod tip through to 10 o'clock (45°) with a definite stop. The effort used should be just enough for the line to become fully extended as it runs out of energy. Lower the rod so that the fly, line, and rod tip all touch down together. Shooting line should be done immediately after the stop by releasing the line from the finger or line hand. False casting is done to work out more line, to dry a fly and to gauge line/length. It is done by performing another back cast after the forward cast without letting the line touch down and then continuing to cast without letting the line touch down. False casts should be kept to a minimum as they may scare fish and result in tangles. Try to slow the casts down to give yourself more time to perform neat casts. If casting into the wind or if there is an obstruction behind, tilt the casting arc forward. There are many common overhead casting faults which include slack line on pickup in which the line may wrap around the rod tip or land all over the caster. The rod tip might be too high to start and may also waggle the wrist, i.e., the rod moves only in an arc. Also creeping the rod forward during the pause phase can also cause the caster to throw the line downwards on the forward cast, i.e., the creep forward results in an arm that is already extended leading to "Tailing Loops."

5.2 Sidearm Cast

This technique is almost the same as overhead cast, however, the differences between the overhead cast and sidearm cast is the rod that is swung on the side of your hand either the left or the right depending on your reel. If the reel or the handle is on the left, then you will swing the fishing rod from the right, and otherwise. However, this technique is often used by highly skilled anglers. Both positive and negative curves can be made. Positive curves are made by over powering the casting while negative curves are made by underpowering the cast. The hard stop helps in positive curve and soft stop in negative curves. Keep the bait close to water so that when it's entering the water it doesn't make a splash. Start with about 2–3 feet of line up the water surface. Now make a circular motion with your hand and stop the line with your thumb as it gets close to the water and bait reaches the fish. Sidearm casting is useful for fishing in windy conditions.

5.3 Double Haul

This cast is designed to increase line speed and shoot line and is great for windy conditions. For this cast, start with your basic fly overhead cast. Just as you feel the rod load, right before you move forward, pull quickly down on the line with your non-dominant hand. This will increase the line speed and further load the rod. Bring the non-dominant hand back to meet your casting hand as the line comes forward. This is a single haul. For the double haul, just repeat this on the back cast with a quick pull at the start and let the hands drift back together as the line unfurls behind you. When you get into the groove, your non-dominant hand will be making a small circle during the cast to control the line tension and make the hauls at the appropriate time.

5.4 Reach Cast

This cast provides an aerial mend which enhances a drag free drift and is useful for fishing in streams. To make this cast, start with the basic cast. Before you lower your rod, allow the rod to shoot a little more line out. As the extra line is coming out, reach the rod out to the desired side. If done properly, the terminal end will be accurate with the fly landing where the rod tip was originally pointed, and just the last

10 feet will have extra slack that will allow for a drift-free float. An upstream reach cast is useful when casting from fast to slower currents.

5.5 Tuck Cast

The tuck cast is a fly fishing essential and it's useful on streamers and wets. Even dry flies get some necessary slack by completing the same motion of a tuck cast on a dry leader. It's a vital fly-fishing tool, not a specialized cast for rare moments. A good tuck cast forces the nymph into the water before the attached leader follows. The nymph may enter at high speed, with enough momentum to dive through the water column and quickly reach the bottom. Or the nymph may enter gently, with less inertia to push it through the water. Hard, soft, and the full range in between—an experienced nymph angler adjusts the speed of the fly's entry, driving the nymph in hard to reach the river bottom faster, or softening the tuck for a shallower, slower sinking effect. The tuck happens when the line unrolls on the forward cast. As the angler stops the rod tip, the nymph trails the leader until the loop unfolds. The nymph reaches the end of the line as it straightens, and when the nymph can't go forward anymore, momentum takes it down into the water. The angler controls the tuck cast by adjusting the speed of the cast and subtlety changing the point where the rod tip stops. The cast is stopped aggressively before it un-loops fully enabling it to sink with a great tension.

5.6 Parachute and Pile Cast

These two similar casts are used in rare circumstances when fishing downstream to rising fish. With the parachute or pile cast, you can create enough slack in the line to drift your fly to the rising downstream fish. To make the parachute cast, use the basic cast but end the forward cast high. Once the fly has landed, lower the rod tip at the current water speed to give the line slack for a drag-free drift. To make the pile cast, start with the basic cast. End slightly higher than average on the forward cast and quickly lower the rod tip. If done properly, a pile of the slack line will be in front of you, allowing the dry fly to drift downstream. This is not an accurate cast, and it is difficult to do in windy conditions.

5.7 Steeple Cast

The steeple cast is useful if there is an obstruction behind you. This is a basic cast but on a different plane. Make the back cast in an upward motion stopping at noon. Once the line unfurls, bring the cast forward, stop at eye level, and lower the rod.

5.8 Roll Cast

The roll cast is a very efficient and effective cast and if one is planning fishing on creeks or streams. The primary purpose of the roll cast is to give the ability to cast when a back-cast is not possible. This may occur, for instance, if there are trees or bushes on either side or behind. The roll cast is also effective when there is a stiff wind at back or need to just quickly reset your fly. In roll cast, we consider the forward strokes rather than pondering what dwells at back. It is highly versatile and can help make casts when in tight quarters. To make this cast, start with the line in front of you on the water. Lift the rod tip smoothly, stopping at about ear level or 11 o'clock. Let the fly line fall behind the rod and make a "D." For the success of this cast, the line must be behind the rod. Once this D has been made, with the rod tip pointing in the direction you want to make the cast, make a powerful forward cast, and use the water surface tension to load the rod and cast your fly.

A roll cast works by using the tension between your fly line and the surface of the water to propel your line forward in a tight loop. This eliminates back cast completely, as well as any false casting. Roll casts are often used in small streams or while nymph fishing. As roll cast uses the tension between your fly-fishing line and the water surface. This tension propels your line forward in a rolling loop until it straightens out and drops onto the water's surface. Some common mistakes during roll cast include lifting the rod too quickly which causes the line to jump off the water and land behind the caster. Failing to form a good "D" loop by not having the rod in the 1 o'clock position also occurs. No pause after lifting the rod and before the forward cast so the line is still travelling back on the water and isn't, therefore, anchored. Applying too much power too soon in the forward cast results in the line slipping on its anchor, which usually results in a cracking noise (from the whip) and line landing loosely on the water. Chopping the rod tip down— rather than driving the rod forward in a straight line to a crisp stop is also important. Creeping the rod forward before the forward cast starts properly—which causes the caster to throw the line downwards on the forward cast, i.e., the creep forward results in an arm that is already extended, so there is nowhere to go but down.

5.9 Snake Cast

The snake cast is like the roll cast. To make this cast, start with the line downstream of you and make a roll with the rod, making a lowercase "c" in the direction of the river. At this point, the line should be in front of you. Now transitioning into a roll cast, the fly should land where you want it to go. This type has lesser risks of tangling.

5.10 Curved Cast

To make a curve cast, simply make your basic cast but do it from a side arm position. The line will hook by overpowering the forward cast, and you will have a curve to your left. You will slice your cast and have a curve to your right by underpowering the cast. Often, by underpowering your cast, accuracy is sacrificed. Bring your basic cast over your non-dominant shoulder and then overpower the forward cast to get a more accurate curve to your right. This will create the curve to the right.

5.11 Bow and Arrow Cast

This is a great cast for small streams where any other cast would be impossible due to bushy overgrowth and other obstacles. Hold the line with your non-dominant hand just above your cork grip to make this cast. Point the rod where you want the fly to go. Pull the line back to your ear and load the rod, much like a bow. Make sure your casting hand holds the line tightly under the index finger against the rod. Then let go of your line in the non-dominant hand (keep the line pinched with the dominant hand), and the rod fling the fly where you want it to go. With this cast, you will not be shooting line. The distance will be dictated by where the line is being pinched against the rod with your index finger.

6 Cast and Retrieve Techniques

Apart from the various casting techniques above, there are several cast and retrieve techniques which are as follows:

6.1 Walking the Dog

In this technique, one uses a top water lure or sub-surface lure that plays under the water surface which gives a twisting action like the appearance of an injured fish and attracts predators. This term is better known as walking the dog. To retrieve the end of the rod tip, point downwards to prevent the bait from jumping up when rolling the line. By rolling a little fast with a certain rhythm, you can also pull the rod occasionally. For some species of pelagic fish, when the water is warm enough, the fish are very active and at that time, the bait that plays on the surface is more quickly grabbed by the fish where the fish hunt for prey near the surface of the water.

6.2 Bottom Bouncing

This technique requires a heavier bait such as a spoon, cast jig, or sinking crankbait so that the bait plays deep near bottom. You need to roll the line by occasionally pulling the fishing rod upwards to make it look like the bait is jumping to attract the fish to grab it. Fishing locations for this technique include swamp area and so on.

6.3 Ripping

In this technique, a bait is thrown to the side of the boat and with a certain speed, the lines are reeled. The most important thing one has to see is the current flowing from the front of the boat to the back. There are techniques depending on the target fish one is fishing for, for example a ferocious fish such as snakehead, mangrove jack, needs a fast retrieving speed. If you are targeting a predatory fish that is a little lazy like Barramundi, then one should retrieve it a little slower. However, the location may not only have one type of fish, there may be several types of fish.

The following retrieve techniques that you can use:

6.4 Slow Retrieve

Slow retrieve is reeling the lines more slowly and as naturally as possible. Lures that are suitable for this retrieve technique are minnow, crankbait, swimbait, and so on.

6.5 Fast Retrieve

A technique where after throwing the lure the lines are reeled quickly so that the bait runs fast on the water. Lures that are suitable for this retrieve technique are minnow, crankbait, swimbait, cast jig, spoon, and so on.

6.6 Stop and Go Retrieve

This technique is a combination technique of slow and fast retrieve, where you play the slow retrieve technique then stop and then play fast retrieve over and over again so that it looks like the bait is injured.

6.7 Cast Popper

This is almost the same technique as the Popping fishing technique, but here a smaller popper or mini popper is used.

7 Precautions to Be Taken While Casting

Casting plugs can hurt but once hooks are added great pain can be inflected if one does not pay attention to where the plug is at all times. Anglers need to be aware of their bait/plug from the time they begin their cast until the time it is retrieved which means looking behind, overhead, and in front. The back cast is the greatest concern. Anglers get so intense about where they are going to cast to that they sometimes forget to look behind them to see if their back cast is going to hit someone or get hung in a tree. When fishing from the bank, someone may come up behind to watch or just be causally walking by. When in a boat, most anglers know their partner is there, but don't realize that their back cast can reach their partner. On the forward cast most anglers are aware of other people, but sometimes forget about power lines and overhanging trees. Once the cast has been made, the plug can get hung up even during practice. One should take precautions about putting too much pressure to retrieve the snagged plug and also while grabbing the line to pull it free. Monofilament line can cut through the skin, so the anglers should wrap the line around the rod handle and use it to pull the line free. Once it comes loose it can shoot back like a bullet and hurt whomever it hits.

8 Conclusion

The diverse terrain and varying water conditions of the Himalayan rivers and streams require anglers to develop a versatile skill set and an in-depth understanding of factors like wind, water currents, and line control. Proper casting form and technique are essential for accessing the most productive fishing spots and presenting the fly in a manner that minimizes disturbance to the fish. Casting is a lifelong pursuit for dedicated fly fishermen, as it allows them to adapt to changing conditions and consistently present the fly in a natural, unobtrusive manner. Through continued practice and refinement of their casting skills, anglers can enhance their chances of success and fully immerse themselves in the thrill and artistry of fly fishing in the Himalayan region. By developing proficiency in casting techniques, sport fishing enthusiasts can not only improve their catch rates but also contribute to the long-term sustainability of these game fish populations through responsible angling practices.

Fly Fishing

11

Abstract

Sport anglers in India, especially those in the Himalayan region, have taken a great interest in fly fishing, a specialized kind of angling. This technique uses a synthetic fly that is cast with a tapered line and a lightweight, flexible fly rod to mimic the natural prey of fish. While fly-fishing seasons vary by region, in the Himalayan rivers the optimum times to fish are from October to November and from mid-February to mid-May. Since anglers must display the fly naturally on the water's surface to tempt the fish, fly fishing is a skill and precision-based sport. To guarantee that these game fish populations remain sustainable, catch-and-release techniques are frequently used. Fly fishing has gained popularity due to its emphasis on technique, relationship with nature, and conservation.

Keywords

Fly fishing · Himalayan region · Fly rod · Artificial fly · Catch-and-release

1 Introduction

Fly fishing is a popular method of angling that involves using an artificial fly, a specialised fishing rod, and a reel to catch fish. It is considered a recreational activity and is often associated with ecotourism because of its focus on natural areas and conservation efforts. Tiny artificial lures that float on or beneath the water's surface are necessary for fly-fishing. As a result, they are too light to be thrown by themselves, as in bait throwing. The key to accuracy and distance in fly casting is a long, light rod and a relatively heavy fly line. Fly-fishing, which was formerly limited to trout and salmon, is now favored by many anglers for black bass, panfish, and numerous other freshwater species. Fly-fishing, once limited to trout and salmon, is

© The Author(s), under exclusive license to Springer Nature Singapore Pte Ltd. 2024
S. T. Mushtaq et al., *Angling in India*,
https://doi.org/10.1007/978-981-97-8876-7_11

now a favorite method for many anglers pursuing panfish, black bass, and a variety of other freshwater and saltwater species that will take an artificial lure.

2 History

Fly fishing is the oldest type of recreational fishing, having its origins in Macedonia circa 200 CE. The earliest accounts of fly-fishing in Europe can be found in the memoirs of English writers from the fifteenth and sixteenth centuries, though the sport had most likely been practiced for at least 200 years before then.

Large, rocky rivers and streams teeming with trout and other species of fish were discovered by immigrants in the United States and Canada. Fly-fishing originated specifically in North America due to the dissimilarities between these rivers and the serene rivers of Europe. By the 1780s, fishing tackle stores in Philadelphia and other cities were advertising a large selection of flies and fly gear. Fly-fishing in North America got its start in the rivers of Newfoundland, Canada; Cape Cod, Massachusetts; and the Cumberland Valley, Pennsylvania. After the railway was built in the middle of the nineteenth century, the Catskill Mountains region of New York became the new fly-fishing epicenter in America because of its many excellent trout rivers. There, early adopters like Edward R. Hewitt and Theodore Gordon helped created an exclusively American school of fly-fishing. Fly-fishing techniques had their own regional variations by the early 1900s in the Pacific Northwest, northern Michigan, and the Rocky Mountains.

3 Tackle

The development of fly-fishing gear has been influenced by technological advancements in the availability of materials. Most of the early rods were constructed from solid wood, mostly willow or hickory. By the mid-1800s, Chinese split bamboo—which offered greater suppleness than wood—was the go-to material for making rods. Bamboo was the preferred material until just after World War II. Then, hollow fiberglass rods that were even lighter took over. A new generation of even lighter and more responsive carbon fiber or graphite rods started to replace fiberglass in the 1970s. Fly fishermen who are always searching for more sensitive tackle continue to favor graphite due to its remarkable strength-to-weight ratio.

The equipment used for fly fishing consists of a fly rod, fly reel, fly line, and leader with a tippet (a thin section of line). Fly rods allow for more accurate casting and delicate presentations because they are often longer and more flexible than traditional fishing rods. The primary function of a fly reel is to retain fly line, but it also aids anglers in battling and landing fish. With great care, the fly line is constructed with a thick, weighted front segment that carries the fly during casting. The leader and tippet, which are tapered sections of monofilament or fluorocarbon line attached to the fly line, allow for a smooth connection between the fly and the fly.

A fly rod's typical length is between 7 and 8 1/2 feet, though shorter and longer sizes are available for particular uses. Fly rods can be made of delicate, expensive split bamboo or strong, reasonably priced fiber glass. Reels fall into two categories: single-action and automatic. The extra line is simply held in reserve by the single-action reels, which can be removed as needed. Often, they have adjustable brakes that can be used to reduce fish runs. In addition to storing line, automatic reels have a coiled spring inside that can be manually wound and activated by a thumb lever to recover slack line or to give and take line to keep it taut while handling the fish.

Fly rods should have a medium action, or curvature, under pressure. It is often described as "parabolic" motion because it is neither too stiff nor too quick. The lines are braided from synthetic threads and coated with a plastic-like substance to give them waterproof, smooth, and heavy qualities. These lines are made to float when fly-fishing dry, or sink when fly-fishing wet. Both types are made level for bait fishing or short casts with flies; double-tapered (narrow at both ends and thus reversible) for longer casts and dry fly fishing; and forward- or torpedo-tapered (more wind resistance) for large or bulky flies, poppers, and bugs. For lines with all these variations, sizes are also prepared.

Tapered monofilament leaders, placed between the line and fly, help the angler cast the fly correctly and conceal the connection from the fish. Leaders normally have a length of 6–12 feet or more, though they can be as long as the rod. The butt diameter should be at least two thirds of the line-end diameter. The size of fly being used determines how fine or coarse the taper should be. If the tip of a leader is too large, a blood knot can be used to attach a finer tippet of monofilament. Artificial flies come in nearly infinite sizes, styles, and variations, but for the novice, a dozen or so well-selected patterns should be plenty.

4 Technique

Using a lightweight fishing rod and feathers, fur, and synthetic materials, a fly fisherman casts an almost weightless artificial fly. Using the fly line's weight to propel the fly forward, the casting technique mimics the movement of insects or baitfish on the water's surface or below. The idea is to present the fly in a way that will entice fish to strike it.

Fly casting involves two basic motions: the forward cast and the back cast. Before beginning the cast, the line is first stretched out for 20–25 feet on the water's surface. The hand that is not holding the rod grabs hold of the line and holds it out to the side between the reel and the first line guide. The rod is raised faster and faster just past the vertical, where the line is stretched almost exactly back. With a flick of the wrist, you can then change the direction of movement, causing the line to curve forward and keep extending until the fly just barely touches the surface. A dry fly is usually thrown upstream and allowed to float with the current until a fish catches it. Other floating flies, like panfish or bass bugs, are cast towards the shore and worked slowly back with short, irregular jerks. A wet fly, or nymph, is an imitation insect larva that can be thrown upstream, downstream, or alongside a river to travel with it or be dragged against it. Underwater lures such as streamers and bucktail flies can be cast slightly upstream, across the stream, or downstream and then quickly moved in the opposite direction to mimic a small fish trying to get away from a larger one.

5 Modern Fly Fishing

Ever since the turn of the twentieth century, fly-fishing has gained popularity. The majority of the sport's fans are female, and they have made many historical contributions. In particular, fly-fishing has been greatly influenced by three American women: Mary Orvis Marbury wrote the first comprehensive book of fly patterns in 1892. Modern fly-tying techniques were created by Helen Shaw in the 1940s and 1950s. One of the world's greatest fly-casters, Joan Salvato Wulff broke many records in the 1950s and 1960s.

Catch-and-release fly-fishing, made popular by Wulff and her well-known fly-fishing husband, Lee Wulff, began among trout anglers in the United States and is now becoming more and more popular worldwide. It can be used to target a wide variety of other species and fishing techniques. Fly fishermen remain at the forefront of global efforts to conserve fisheries because of their involvement in conservation organizations.

Since fly-fishing was originally developed as a method of fishing primarily for trout and salmon, it has expanded to include a wide range of species. Freshwater fly fishermen catch panfish, pike, bass, and other species. Many fishing organizations sponsor annual fly-casting competitions for accuracy and distance. The method of "shooting the line," which involves the angler unreeling a significant amount of line and holding it coiled in his free hand prior to the cast, is frequently used to achieve long casts.

6 Types of Flies

Understanding the different types of flies used in fly fishing is crucial for any angler. Each type of fly is designed to imitate various stages of an insect's life cycle or other prey items in the fish's diet.

6.1 Dry Flies

These flies are designed to float on the water's surface, mimicking adult insects such as mayflies, caddisflies, and stoneflies (Fig. 11.1). Dry fly fishing is often considered the pinnacle of the sport due to the visual aspect of watching a fish rise to take the fly. Popular dry fly patterns include the Adams, Elk Hair Caddis, and Royal Wulff.

Fig. 11.1 Parachute Adams dry fly. Freyfisher, CC BY-SA 2.0. https://creativecommons.org/licenses/by-sa/2.0, via Wikimedia Commons

Fig. 11.2 Jessy Hard wet fly (Picture Credit : Ubaid Wani)

6.2 Wet Flies

Wet flies are designed to sink below the surface and imitate insects in their underwater stages, like nymphs or drowned adults (Fig. 11.2). These flies often have soft, flowing materials that provide lifelike movement underwater. Classic wet fly patterns include the Partridge and Orange, and the Soft Hackle.

6.3 Nymphs

Nymphs represent the juvenile stages of aquatic insects and are fished below the surface. They are highly effective since fish feed predominantly underwater. Common nymph patterns include the Pheasant Tail, Hare's Ear, and Prince Nymph.

6.4 Streamers

Streamers are larger flies that mimic baitfish, leeches, or other sizable prey. These flies are often fished with an active retrieve to elicit aggressive strikes from predatory fish. Popular streamer patterns include the Woolly Bugger, Muddler Minnow, and Clouser Minnow.

6.5 Terrestrials

These flies mimic land-based insects like ants, grasshoppers, and beetles that accidentally fall into the water. Terrestrials are especially effective during the summer months. Examples of terrestrial patterns include the Foam Beetle, Grasshopper, and Chernobyl Ant.

7 Fly Tying

Fly tying is the art of creating artificial flies by combining various materials onto a hook. This skill allows anglers to customize their flies to match local insect hatches and specific fishing conditions. The materials used in fly tying can be natural or synthetic. Common natural materials include feathers (such as hackle, marabou, and pheasant tail), fur (rabbit, deer hair), and various types of animal hairs. Synthetic materials, like foam, rubber legs, and synthetic fibers, are also widely used to enhance durability and buoyancy.

8 Fly Fishing for Different Species

Different species require different approaches and techniques in fly fishing.

8.1 Trout

Fly fishing for trout often involves matching the hatch, which means selecting flies that closely resemble the insects currently emerging in the area. Techniques include dry fly fishing, nymphing, and streamer fishing. Key considerations are water temperature, insect activity, and trout behavior (Fig. 11.3).

Fig. 11.3 Red tag trout wet fly. Mike cline, CC BY-SA 4.0. https://creativecommons.org/licenses/by-sa/4.0, via Wikimedia Commons

8.2 Salmon

Salmon fishing typically takes place in larger rivers and streams. Anglers use larger flies, often brightly colored, to attract the attention of migratory salmon. Techniques include swinging flies downstream, dead drifting, and stripping streamers (Fig. 11.4).

8.3 Bass

Fly fishing for bass can be highly exciting, as bass are aggressive predators. Poppers, streamers, and large terrestrial flies are effective. Techniques involve casting towards structures like logs, weed beds, and rock piles, and using an active retrieve to provoke strikes.

8.4 Saltwater Species

Saltwater fly fishing targets species such as bonefish, tarpon, and striped bass. These fish often require longer casts and more robust gear. Techniques include sight fishing for bonefish on flats, using large streamers for tarpon, and casting into surf or around structures for striped bass.

Fig. 11.4 Durham Ranger Salmon fly. MichaelMaggs, CC BY-SA 3.0. https://creativecommons.org/licenses/by-sa/3.0, via Wikimedia Commons

9 Conclusion

Fly fishing, with its rich history and technical advancements, has transformed from a niche pursuit for trout and salmon to a widely practiced sport targeting various fish species across diverse environments. Technological innovations in tackle and gear have enhanced the fly fishing experience, allowing for more precise and effective casting. The sport's techniques, focusing on mimicking natural insect behaviors, require skill and practice, making it a challenging yet rewarding activity. The contributions of pioneering anglers, especially women, have significantly shaped modern fly fishing practices. Today, fly fishing continues to thrive as both a recreational and conservation-oriented endeavor, fostering a deep connection with nature and promoting sustainable fishing practices worldwide.

Flipping

12

Abstract

Flipping is a specialized fishing method that's utilized in tight or heavily covered settings, including dense vegetation, under overhanging trees, or around docks, to offer bait properly and quickly. This is a popular approach for bass fishing where you pull a line length from the spool with your non-dominant hand and swing the rod to accurately put the bait. It gives fishermen more opportunities to catch fish by allowing them to target fish that are lurking in hard-to-reach areas. Specialized gear is needed for flipping, including as thick braided lines, fast bait-casting reels, and heavy-duty rods. The method uses a variety of techniques, including pitching, skipping, lowering, dabbing, and punching, to offer lures in a variety of settings. To maximize effectiveness and success, choosing the right cover and knowing when to flip are essential. Although flipping has several benefits, like accurate casting, little disturbance, and adaptability, it also requires training and the appropriate tools to become proficient.

Keywords

Flipping · Baitcasting · Braided lines · Bass fishing

1 Introduction

In sport fishing, the term "flipping" describes a method of presenting bait or lures precisely and accurately, especially in confined spaces like around docks, under overhanging trees, or in dense vegetation. In this technique, one uses the non-dominant hand to pull an arm's length of line off of the reel. Then the rod is used to swing the bait into close proximity areas. It is a fast and efficient way to get baits in front of fish that make their homes in thick cover that is avoided by most finesse fisherman. In other words, flipping is the art of using the momentum of the bait and

© The Author(s), under exclusive license to Springer Nature Singapore Pte Ltd. 2024
S. T. Mushtaq et al., *Angling in India*,
https://doi.org/10.1007/978-981-97-8876-7_12

the swing of the rod, coupled with an arm's length of line in non-dominant hand to place a bait in close proximity to the area of interest. It is a method for presenting bait accurately and efficiently in close quarters. This technique keeps the bait low to the water's surface, which allows to conveniently place lures in and around shallow cover. Flipping consists of small accurate casts when fishing shallow structures. These short, accurate casts allow the angler to put bait into small holes or gaps in the structure being used for fishing. Although this method is frequently used for bass fishing, it can also be employed for many other kinds of fishing. It is a flexible and efficient method that lets fishermen increase their chances of success in a variety of fishing conditions by focusing on fish in difficult or difficult-to-reach places.

2 Equipment Used for Flipping

2.1 Rods and Reels

For flipping, anglers typically need specialized equipment, such as heavier, shorter fishing rods with a strong backbone that are sensitive (generally 7–8 feet long). These rods are used in conjunction with heavy-line baitcasting reels that have a large capacity and are spooled with braided lines, which provide increased sensitivity and control. Longer rods (usually 7–8 feet) allow for longer pitches and better leverage when flipping (Fig. 12.1). To extract fish from deep cover, medium-heavy to heavy power rods work best because they can support the weight of larger baits.

Fig. 12.1 Fly rod with reel (Picture credit: Dr. Musadiq Allaqaband)

For their sensitivity and speedy hook-setting abilities, fast action rods which bend more at the tip are recommended.

The reels used for flipping should be of high speed and be able to pick up line quickly when a fish inhales the bait for a solid hookset. Some popular reels come with a "flipping switch" which, when engaged, does not allow the thumb bar to lock in free spool, but as soon as you remove your thumb the reel will instantly engage the drive train. This allows for instant hooksets if a fish eats the lure on the fall and you don't have to switch hands and crank the reel to engage before you can set the hook. To handle larger fish's unexpected runs, reel that has a strong and smooth drag system is preferred. Reels with a high gear ratio such as 7:1 or 8:1 are popular among anglers for quick retrieval of fish from cover once they have been hooked. Because of their accuracy and power, baitcasting reels are the preferred option.

A baitcasting reel is a type of fishing reel that is popular among experienced anglers for its advanced design and functionality. It features a spool that rotates during casting, allowing for precise lure placement and control. While it offers greater power and accuracy, mastering its operation typically requires practice due to its more intricate mechanics compared to other types of reels like spin cast or spinning reels.

2.1.1 Casting Mechanism of a Baitcasting Reel

2.1.1.1 Spool Tension Control

Before casting, anglers adjust the spool tension control knob located on the side of the reel. This knob regulates the speed at which the spool rotates during casting, helping to prevent backlash.

2.1.1.2 Thumb Control

Unlike spinning reels where the line is released by opening the bail, in baitcasting reels, the angler uses their thumb to control the line. As the lure is cast, the angler applies pressure on the spool with their thumb to prevent it from over-spinning.

2.1.1.3 Casting Motion

To cast, the angler holds the rod with one hand and depresses the reel's thumb release button or lever with their thumb using the other hand. This disengages the spool, allowing it to rotate freely.

2.1.1.4 Forward Motion

With a smooth forward motion of the rod, the angler propels the lure towards the desired target. As the lure travels through the air, the spool spins, releasing line.

2.1.1.5 Thumb Braking

Throughout the casting motion, the angler monitors the speed of the spool with their thumb, applying pressure to slow it down as needed to prevent backlash. This requires a delicate balance of releasing the right amount of line while maintaining control over the spool's rotation.

2.1.1.6 Line Release

Once the lure reaches the desired distance, the angler stops the forward motion of the rod and applies pressure to the spool with their thumb to halt the line's release.

2.1.1.7 Engagement

After the cast is completed, the angler engages the reel by releasing the thumb pressure and engaging the reel's braking system, which helps to control the spool's rotation and prevent overruns.

2.2 Hooks, Weights, and Lines

It is recommended to use hooks that are heavier gauge wire with a straight shank. These features allow less flex in the hook on the hookset, and more force directed to the hook's point. The Trokar Flippin hook is considered by many anglers as best of them. Common options include heavy-duty straight shank hooks, flipping hooks, and offset worm hooks. To help the bait or lure drop fast and pierce tight cover, tungsten or lead weights, also referred to as "flipping weights" or "bullet weights," are placed to the fishing line above the hook.

The lines used for these techniques should be strong and tough and frequently checked for nicks and scratches throughout the day. Fishing lines that have been braided are often employed for flipping due to their small diameter, great strength, and low stretch. Because braided lines are so sensitive, fishermen may feel even the smallest bites through thick vegetation. Additionally, they offer the strength required to remove fish without breaking out of dense vegetation or structural elements. Depending on the target species and fishing conditions, anglers usually utilize strong test braided lines, weighing over 50 pounds. Abrasion resistance of the braid is critical for fighting fish in cover and gives a better success rate when dealing with submerged vegetation or docks.

2.3 Lures

Lures should be fast sinking with plenty of movement (plastic jig or soft plastic).

2.3.1 Jigs

Flipping jigs are specifically designed to be sturdy and tough in difficult conditions. They can go into heavy wood and brush without getting damaged, making it possible to fish in heavy cover. The head of the jig should be very compact, ideally with a recessed line tie. Small flipping or pitching jigs that imitate crawfish or baitfish quite well, especially when combined with a soft plastic trailer.

Flipping jigs must be designed to prevent hanging up, so the weed guard tends to be firmer than most other jig types. Rattles are a great addition when flipping a jig and the best models will come with one already integrated. Chunks and other creatures are among the best trailers to use on a flipping jig. Many flipping jigs come with a chip-resistant head and this is always a good option to go for. The jig is likely to hit hard rocks and sand along the way, so having a chip-resistant option is always good. There are a number of head designs available but stand-up flipping jigs are often the best. They will sit on the bottom of the water nicely and they give the appearance of a crayfish to any passing bass. Finding the right hook is also incredibly important when choosing a flipping jig. Wide-gapped hooks are best because they improve the angle of the strike, ensuring that you penetrate deep into the jaw of the bass.

2.3.2 Artificial Bait

There are basically two types of presentations, horizontal and vertical. Horizontal presentations, such as spinner baits (Fig. 12.2) and crank baits are mostly used to cover water and find more aggressive or active fish. Creatures, craws, and worms are popular choices, as they mimic natural prey found in cover. Vertical presentations like flipping is used to present lures to somewhat neutral or inactive fish. These vertical presentations must be learned in order to consistently put you in the position to catch fish. To flip a bait, start by taking the rod in your dominant hand and let out about 15 feet of line. With your opposite hand, grasp the line between the reel and the first line guide. Swing the rod toward the target while releasing the excess line with your other hand. Your lure should land softly, barely making a splash.

Fig. 12.2 Spinner bait. https://upload.wikimedia.org/wikipedia/commons/f/fa/Spinnerbait.jpg

2.4 Protective Gears

Flipping require expertise and technique, and as such having the proper equipment tailored to these methods will greatly increase efficiency and success rate. Gloves provides the increased grip, particularly while flipping for extended periods of time. Eye Protection is also necessary during flipping. Therefore polarized sunglasses help shield our eyes from sharp hooks or weights while also aiding in fish identification.

3 Techniques of Flipping

3.1 Pitching

Pitching is a technique wherein short, precise casts are made to nearby targets. To precisely swing the bait or lure underhand towards the target, anglers use their wrists. It's a low-key method that reduces noise and works well when fishing near docks, overhanging trees, or other cover.

3.2 Skipping

Skipping is a technique where anglers skip their bait or lure across the water's surface to reach targets underneath overhangs, docks, or other structures. It requires a sidearm or underhand casting motion and precise timing to achieve the desired trajectory. Skipping allows anglers to access hard-to-reach areas where fish may be hiding.

3.3 Dropping

In order to target fish that are holding close to the bottom or structure, dropping entails lowering the bait or lure straight into the water, usually from a boat or kayak. To regulate the bait or lure's fall and maintain it in the strike zone for extended periods of time, anglers utilize their rod. Fishing vertical cover such as brush piles, bridge pilings, or submerged trees works well with dropping.

3.4 Dabbing

Dabbing is a finesse technique where anglers gently present their bait or lure to fish holding in shallow water or around cover. It involves lightly tapping or dabbing the bait on the water's surface to mimic natural movements and entice bites from wary fish. Dabbing is often used in clear water or when fish are in a neutral or negative feeding mood.

3.5 Punching

Punching is a power fishing method that allows anglers to get to fish that are lurking beneath dense undergrowth. Fishing enthusiasts utilize hefty weights, commonly called "punch rigs," to penetrate plant mats such as matted grass or lily pads. The bait or lure is carried through the vegetation by the weight, falling straight to the fish below.

4 Identifying Appropriate Covers and Time for Flipping

Flipping is a fishing method intended to discreetly present baits in spots where bass are probably hiding, therefore identifying and comprehending the appropriate covers is essential for success. Flipping is most effective whenever bass are holding on shallow, heavy cover like grass, laydowns, or docks. The best places to flip and pitch are wherever a traditional cast won't allow you to accurately or efficiently place your bait (Table 12.1).

Ideal spots are around flooded bushes, overhanging branches, laydowns, grass beds, tule lines, and even undercut banks. It's difficult to place your lure in a small, basketball sized hole in a grass bed with a normal cast, but by learning to flip and pitch, it's easy for even novice anglers. Depending on the lake, this may only be during a portion of the year, but even the deepest lakes are likely to harbor at least some shallow bass all year round that can be caught by flipping.

Spring tends to be the best time for pitching and flipping, as the fish move from their deeper winter holds to the shallows to spawn. Summer pitching will be largely dependent on water temperature. If water temperatures rise too much, bigger fish will seek refuge in deeper, more oxygenated water. Fall can also be an excellent

Table 12.1 Common bass fishing covers and their descriptions

Cover	Description
Fallen trees and logs	Sunken or partially submerged wood. These structures offer shelter and draw in smaller prey, making them prime locations for bass
Dock pilings	Wood or metal frameworks supporting docks. Bass utilize these for protection and as vantage points for ambushing prey
Lily pads	Floating green pads, typically with stems below the water. Bass frequently hide under lily pads, taking advantage of the shade and using them as strategic ambush spots
Grass mats and vegetation	Dense surface or underwater vegetation. Thick vegetation provides bass with excellent cover and an abundant hunting area
Rock piles and boulders	Clusters of rocks or large stones in the water. They offer structure for bass and often attract crayfish, which are a favored prey for bass
Reed and cattail beds	Tall, slim vegetation rising from the water. Offering ample coverage, they are often abundant with prey such as frogs and insects, enticing bass
Brush piles	Clusters of branches and brush submerged in the water. They form intricate structures that bass utilize as hiding spots

season to try flipping, as bass will stalk the shallow areas, looking to eat as much as possible before the temperatures drop.

5 Advantages and Disadvantages of Baitcasting Reels

Baitcasting reels offer precise casting control, allowing anglers to target specific spots with accuracy, especially in challenging conditions like heavy cover or windy weather. They provide more power and torque, making them suitable for catching larger fish species. Baitcasters can handle a wide range of fishing techniques, including flipping, pitching, and casting heavy lures. Many baitcasting reels come with adjustable settings such as braking systems and drag, allowing anglers to fine-tune their setup according to their preferences and fishing conditions. High-quality baitcasting reels are built to withstand heavy use and are often more durable than other reel types.

Baitcasting reels have a steeper learning curve compared to spincast or spinning reels. Beginners may experience backlash or tangles until they become proficient at casting. Quality baitcasting reels can be more expensive than other reel types, making them less accessible to budget-conscious anglers. Baitcasters require regular maintenance, including cleaning and lubrication, to ensure smooth operation and prevent issues like corrosion. Improper spool tension or casting technique can result in backlash, where the fishing line tangles on the spool, requiring time-consuming untangling. Baitcasting reels may not perform as well with lightweight lures or fishing lines, as they are designed for heavier applications.

6 Advantages and Disadvantages of Braided Fishing Lines

Braided lines are known for their high tensile strength, which means they can handle larger fish without breaking. They are made from materials like Spectra or Dyneema, which are extremely strong and resistant to abrasion. Compared to monofilament and fluorocarbon lines of the same strength, braided lines have a much thinner diameter. This allows anglers to spool more line onto their reels and achieve greater casting distances. The lack of stretch in braided lines makes them highly sensitive to bites and underwater structures. This feature helps anglers detect even the slightest nibbles and improves hook-setting efficiency. Braided lines are highly resistant to UV rays and water absorption, making them more durable over time. They do not degrade as quickly as monofilament lines, providing long-term use.

Braided lines are more visible in clear water, which can sometimes spook fish. This can be mitigated by using a fluorocarbon leader, which is less visible underwater. These lines can be more challenging to knot properly and can be prone to slipping if not tied with the correct knots. They also can dig into the reel spool under heavy pressure, making them harder to handle. Braided lines are generally more expensive than monofilament and fluorocarbon lines. However, their durability often makes up for the higher initial cost in the long run.

7 Advantages of Flipping

7.1 Precision Casting

Flipping allows anglers to precisely present baits or lures into tight spaces such as under docks, around vegetation, or into pockets within structure. This precision casting increases the likelihood of placing the bait directly in front of fish without spooking them, maximizing the chances of a strike.

7.2 Minimal Disturbance

Unlike traditional casting techniques that may create splashes or disturbances on the water's surface, flipping produces minimal disturbance. This subtle presentation is especially advantageous in clear or calm water conditions, where fish may be easily spooked by noise or disturbances.

7.3 Targeting Hard-to-Reach Areas

Flipping enables anglers to reach areas that are difficult or impossible to access with other casting techniques. By presenting baits or lures in areas with heavy cover, such as under overhanging trees or deep within vegetation, anglers can target fish that are sheltering or feeding in these prime locations.

7.4 Increased Hookup Ratio

The short distance between the angler and the target area during flipping allows for quick hook sets. This proximity, combined with the sensitivity of the gear and line used in flipping, results in a higher hookup ratio. Anglers can detect bites more easily and react quickly to set the hook, increasing their chances of landing fish.

7.5 Versatility

Flipping can be adapted to various fishing environments and target species. Whether fishing for bass in freshwater lakes, targeting snook in saltwater mangroves, or pursuing panfish in shallow ponds, flipping can be effective in a wide range of scenarios.

7.6 Reduced Snagging

Flipping with weedless or snag-resistant baits reduces the likelihood of getting hung up on underwater obstacles such as rocks, logs, or vegetation. This helps anglers spend more time fishing and less time dealing with frustrating snags and lost gear.

7.7 Efficiency

Flipping allows anglers to cover a lot of water efficiently, especially when working methodically through areas with dense cover or structure. By systematically presenting baits or lures in key target zones, anglers can maximize their time on the water and increase their chances of success.

Overall, flipping is a versatile and effective technique that offers anglers a strategic advantage when targeting fish in challenging or hard-to-reach areas. Its precision, minimal disturbance, and ability to target prime fishing spots make it a valuable tool in the angler's arsenal.

8 Conclusion

In sport fishing, flipping is a crucial tactic, especially when pursuing fish in regions with thick cover or difficult access. Anglers can increase their odds of good strikes by learning this technique, which will allow them to offer their baits with precision and minimal disturbance. To flip a bait properly, specialized gear is needed, like thick braided lines and high-speed baitcasting reels with heavy-duty rods. Pitch, skip, drop, dab, and punch techniques are some of the more adaptable methods for a variety of fishing situations and settings. Fishing results can be greatly improved by knowing the right covers to use and when to flip. Flipping is a beneficial technique for any professional angler, despite the fact that it needs talent and the proper equipment. Its advantages include reduced snagging, higher hookup rates, and efficiency.

Pitching

13

Abstract

Pitching is a specialist casting method that is heavily utilized, especially in places with extensive cover or clear, shallow waters where accurate lure placement is essential. This technique minimizes disturbance to the surrounding environment by carefully and silently sending lures into tight areas, like beneath docks or among plants. By lowering fish attentiveness and increasing catch rates, effective pitching improves angler efficiency. This is especially true when paired with moral fishing methods like catch-and-release. This chapter examines the background, method details, tackle needs, essential components, advantages, and disadvantages of pitching for recreational fishing, with a focus on how it can improve fishing success while having the least negative environmental impact.

Keywords

Techniques for pitching · Bass fishing · Casting · Placing lures · Precision fishing · Angling

1 Introduction

Pitching is a casting technique in fishing that minimizes disturbance to the lure as it enters the water. Although it can be used in other fishing situations, it is frequently utilized when bass fishing. It is especially useful when fishing in clear, shallow water or when fish are easily startled, both of which call for subtle casting. Additionally, pitching can increase accuracy, enabling fishermen to precisely position the lure. It enables fishermen to set their lures in confined spaces where fish are probably hiding or looking for cover, like beneath docks, among vegetation, or close to cover.

© The Author(s), under exclusive license to Springer Nature Singapore Pte Ltd. 2024
S. T. Mushtaq et al., *Angling in India*,
https://doi.org/10.1007/978-981-97-8876-7_13

For catching fish that are located at a short to medium distance, it is a well-liked and effective method. Anglers should aim their rod at the intended target while pitching, letting the bait rest in their hand. They should then make a devious "pitch" to their target using their rod. Anglers usually target areas 20–40 feet out from the boat when pitching. The likelihood of upsetting the fish or making them defensive decreases when anglers are able to place their bait discreetly and precisely. Pitching is a technique that can reduce fish predator alertness, increasing angling efficiency. When combined with other ethical fishing methods, proper pitching can dramatically raise fish survival rates after release.

2 History

There is limited historical documentation on the practice of pitching as a specialized fishing technique. The need to place lures in confined spaces, such as dense cover, prompted its development. This method proved more effective than traditional casting, particularly in environments requiring precise lure placement. Tommy Biffle, an angler from Oklahoma, is credited with the refinement and popularization of this technique in the mid-1970s. Biffle, upon receiving a Fenwick Flippin' Stick via mail, found it to be stiff, resembling a pool cue, and weighing a substantial amount. Despite this, he quickly mastered flip-casting and utilized this method to secure numerous tournament wins. His approach to flip-casting involved "just kind of swinging it out there" until he achieved a flip. This method allowed him to maintain the jig in the water longer, increasing his chances of catching bass. Biffle's proficiency in this technique provided him with a significant advantage, enabling him to cast farther from the fish and the cover. He could also effortlessly perform various pitches, reaching targets beyond forty feet. This gave him a competitive edge over traditional flipping methods.

In the initial days of its use, Biffle could achieve a day's worth of fishing with ease, trying different pitches until he succeeded in reaching and catching fish. Moreover, his ability to handle heavier gear without difficulty contributed to his success. Prior to competing in Bassmaster tournaments, Biffle participated in various local events that attracted prominent anglers. His involvement in the sport led to broader recognition and adoption of his methods. During a notable pairing with Rick Clunn at a Lake of the Ozarks tournament in early 1985, Biffle's technique played a pivotal role. With three of Clunn's four Bassmaster Classic victories already secured, Biffle's approach to flipping proved instrumental in their success. While Clunn focused on flipping to buck bushes, Biffle effectively pitched to areas that Clunn's flipping could not reach. This strategy consistently resulted in significant catches, underscoring the effectiveness of Biffle's method.

3 Technique

When the target cover is within one or two rod lengths away, pitching is used to drop a bait straight into any openings in the cover, like a partially submerged branch. To make an underhand cast, hold the fishing rod with the reel facing up, and extend your arm. This is known as the pitching technique. The angler's goal is to quietly and precisely deliver the lure to the intended target area. The arm is swung smoothly, the wrist is flicked slightly, and the lure is released at the appropriate time.

4 Tackle

Most anglers choose slightly shorter rods for pitching, usually 7 feet, with the most common range being 6′8″–7′3″. The shorter length will facilitate the technique's required underhanded delivery. The ideal rod should have a flexible tip with plenty of backbone to draw large fish out of dense cover and deliver the bait as silently and precisely as possible. The reels used for pitching can be the same reels used for other techniques, but they should spin at a high speed (6:1) to ensure the big bass can't turn around in the thick cover and wrap itself up, making it impossible for you to retrieve it, and to pick up line quickly when a fish inhales your bait for a strong hookset.

Anglers prefer jigs that most often imitate the crayfish, which is represented by soft plastic or realistic-looking crawdad baits, when fishing for species like bass. Nevertheless, many flip baits don't seem to have any natural equivalents. Anything that can be inserted into the dense cover without fouling often looks like it will be able to catch fish. However, some soft plastic baits are ineffective because of the design; lures with long, ribbon-like arms and tails, for instance, can tangle in tules, reeds, or branches and miss their target fish's habitat.

Large, thick, sharp wire hooks, long skirts, sharply pointed noses, and robust weedguards are features found on most pitching jigs. Anglers who use soft plastic lures spend money on premium hooks. Most anglers peg the weight when Texas-rigging their flip baits to avoid having the bait and weight land on opposite sides of a branch, which would keep the bait from reaching the fish.

5 Key Elements of Pitching

5.1 Accuracy

Anglers can target specific areas where fish are likely to be present by using pitching to precisely place lures. Fishermen can improve their chances of attracting a strike by rigging the lure near cover, beneath docks, or along the edges of structures.

5.2 Stealth

To avoid frightening fish, pitching entails introducing the lure into the water gradually and deliberately. The goal of anglers is to cause as little disruption as possible while emulating the local prey's natural movements.

5.3 Line Control

When pitching, it's important to keep the fishing line under control. Anglers can control the cast distance, manage the amount of line released, and avoid line backlash by using their thumb on the reel's spool.

5.4 Lure Selection

Depending on the type of fishing and the species being pursued, several lures can be used for pitching. Jigs, soft plastics, creature baits, and other lures that imitate local natural prey are popular choices.

6 Benefits of Pitching

Pitching offers several advantages for anglers:

6.1 Targeting Fish in Tight Spots

By pitching, fishermen can place lures where fish hide, increasing the possibility that a strike will occur.

6.2 Precision

Lures can be precisely positioned to attract fish in particular areas, like under docks, around vegetation, or close to the edges of structures.

6.3 Avoiding Obstacles

Anglers can reduce the likelihood of becoming entangled or snagged in dense cover by putting lures directly in the areas they want to target.

6.4 Increased Stealth

Pitching involves a deliberate and delicate introduction of the lure into the water, which lowers the possibility of surprising fish and increases the likelihood of a successful presentation.

7 Disadvantages

While pitching has its benefits, it also has some disadvantages:

7.1 Skill Requirement

Pitching requires a high level of skill and practice to master. Beginners may find it challenging to achieve the necessary accuracy and stealth.

7.2 Limited Distance

Pitching is effective at short to medium distances, typically within 20 to 40 feet. This can be a limitation when fish are located farther away.

7.3 Equipment Sensitivity

The technique requires specific equipment, such as shorter rods and high-speed reels. Using improper equipment can reduce effectiveness and increase frustration.

7.4 Physical Demand

Repeated pitching can be physically demanding on the arm and wrist, especially during extended fishing sessions.

7.5 Environmental Constraints

Pitching is most effective in certain environments, such as those with lots of cover or structures. It may not be as useful in open water or areas with sparse cover.

7.6 Potential for Snags

Despite its precision, pitching into dense cover increases the risk of snags, which can result in lost lures and wasted time.

8 Conclusion

Pitching is a skill that takes experience and repetition to become proficient in. It has long been an angler's preferred method. However, practice is necessary. In addition to honing their accuracy and timing, anglers must gain a feel for controlling the rod, reel, and line. Anglers can increase their overall success on the water and effectively target fish in particular areas by improving their pitching skills.

Rigging

14

Abstract

This chapter focuses into the intricacies of fishing rigs, essential tools for anglers across diverse fishing environments. It comprehensively covers the components and construction techniques crucial for effective rigging, including leaders, knots, hooks, and sinkers. Each component's selection and arrangement are tailored to target species and fishing conditions, ensuring optimal performance. Techniques such as knot tying, adjusting leader lengths, and selecting appropriate tackle are detailed to enhance rigging proficiency. Specialized rigs like trolling rigs, bottom fishing rigs, and live bait rigs are explored in depth, highlighting their specific applications and advantages. Whether in freshwater, saltwater, or surf fishing scenarios, understanding and mastering various rig types significantly elevate anglers' success rates. This chapter serves as a practical guide for anglers of all levels seeking to refine their rigging skills and maximize their fishing experience.

Keywords

Rigs · Anglers · Knot · Leader length · Trolling

1 Introduction

A fishing rig is a configuration of equipment used for fishing. It defines how leaders, sinkers, bobbers, lures, swivels, and dodgers can be attached to a length of fishing line. It can be put together using a combination of lines, hooks, sinkers, bobbers, swivels, lures, beads, and other fishing gear. It can be carried by hand, attached to a boat or pier, or held in place by a rod. Some rigs are made to float near the water's surface, while others are made to sink to the bottom. Many rigs are made

© The Author(s), under exclusive license to Springer Nature Singapore Pte Ltd. 2024
S. T. Mushtaq et al., *Angling in India*,
https://doi.org/10.1007/978-981-97-8876-7_14

specifically for catching a particular species of fish, but they can also be used for a variety of fish species.

2 Parts of a Rig

A rig is a group of hooks, sinkers, swivels, and leader lines that are attached to the end of a fishing line or terminal tackle. Since fishing methods differ depending on the kind of fish, the environment, and other elements, rigs are designed to fit the situations in which they are employed. Snap swivels are a common means to make rigs detachable (though not necessarily necessary) from mainline, which helps fishermen swap up rigs rapidly in the event that fishing conditions change. A rig's standard components are the leader, the joints (fastened with knots or swivels), the hook (or lure), and the weights. A leader line is a fishing line that joins the various parts of the rig together. Typically, a monofilament line serves as the leader because of its poor visibility in water (as compared to braided line for example) and its elastic nature provides some shock absorbance against strong takes from fishes.

2.1 Leader Line

The objective of the mainline is to have the lowest possible line diameter and the highest possible strength so that you may use lighter sinkers and load more lines into your fishing spool or reel with less drag against the current. In contrast, leader lines endure a variety of penalties from fish and other items they come into contact with, much like front-line soldiers. Since the leader (or rig) is frequently closest to causes of abrasion, such as rocks or other objects in the water and the teeth of fish, it should normally be more abrasion resistant than the mainline. Therefore, big diameter lines or lines built specifically with abrasion resistance make good options for leader lines. In situations where water visibility is high or where the line is intended for fish species that are easily spooked, lighter leader lines or fluorocarbon monofilament lines will work better.

2.2 Joints (Knots and Swivels)

To connect lines to the mainline or to other parts of the terminal tackles, swivels and knots are utilized. Any fishing approach nearly always involves the usage of knots, so it's helpful to be able to tie a few common knots quickly and correctly when one is out fishing. The majority of the time, a rig breaks at or close to a knot. Good knots will usually keep 60–90% of their line strength, even though knots often weaken line strength. With some knots, one might theoretically maintain 100% line strength. However, they are dependent on the type of line, line diameter(s), and knotting technique, much like any other knots. On the other side, a badly made knot can cause the line strength to drop as low as 20%. To preserve the greatest amount of line strength, knotting requires extra care and attention. Whenever possible, moisten the knot with water before tightening it.

Swivels are used in a similar manner as knots. Because they can rotate, swivels have the advantage of easing line twists. Convenience is another benefit; snap swivels' ability to open and shut a loop makes switching out rigs or artificial lures simple. Most swivels have a rating for strength. Since most swivels have a strength rating, the one that best suits the overall strength of the rig should be selected. High-quality swivels should be bought if one must utilize incredibly tiny ones.

2.3 Choice of Hooks

Two primary functions of a hook are to pierce the flesh of the fish and to hold it in place until it is landed. Therefore, when selecting hooks, consideration must be given to the size, strength, and sharpness of the hook. There are numerous varieties of hooks, each designed to be used for a particular kind of fish and in a variety of conditions. The size of the hooks chosen should correspond to the size of the intended fish. It is not suggested to use small hooks, even those of exceptional quality and strength, on larger fish. This is because smaller hooks are more likely to pull off flesh more readily than larger ones since they grasp less of the flesh of the area around the mouth of the fish.

Hook strength is influenced by the material composition of the hook. Hooks with a high steel content are often pliable, and as the carbon composite content rises, the hook's hardness increases. While high diameter hooks are harder and less prone to break when fighting a large fish, low diameter hooks offer the advantage of having a high penetration rate.

2.4 Fishing Weights (Sinkers)

Weights or sinkers assist in casting the rig to the intended spot and in maintaining the rig's position in the water. Depending on the conditions in which a rig will be used, weights can be added to different areas of the rig. For example, a basic weight at the bottom can serve as an anchor, or weights can be added to other parts of a rig to create desired action or position in bottom, midwater, or surface rigs. Larger sinkers are needed in high current situations in order to lower the rig to the seafloor or any other desired location.

3 Essential Rigging Techniques

Rigging techniques are fundamental skills that every angler should master to optimize their success on the water. Whether you're targeting freshwater bass, saltwater snapper, or anything in between, the way you rig your tackle can significantly impact your ability to attract, hook, and land fish.

3.1 Tying Knots

Mastering a few essential fishing knots is essential for rigging success. Popular knots include the Palomar knot for attaching hooks and the Improved Clinch knot for securing line to terminal tackle.

3.2 Adjusting Leader Length

The length of your leader can significantly impact your rig's effectiveness. Short leaders are ideal for presenting bait or lures close to the main line, while longer leaders allow for greater freedom of movement and can be useful in clear water or when targeting wary fish.

3.3 Balancing Tackle

Properly balancing your tackle involves selecting the right combination of line, leader, hooks, and weights to match the size and behavior of target fish. Overly heavy or bulky tackle can spook fish, while tackle that's too light may not hold up to the rigors of a spirited fight.

3.4 Fine-Tuning Bait Presentation

Experimenting with different bait presentations can help you dial in the perfect setup for enticing fish. Techniques such as free-lining, bottom fishing, and trolling each have their own unique applications and require careful attention to detail.

4 Specialized Rigging Techniques

4.1 Trolling Rigs

Trolling involves dragging bait or lures behind a moving boat to cover a large area of water and entice fish that are actively feeding or cruising. Specialized trolling rigs, such as spreader bars and daisy chains, are designed to simulate a school of baitfish and can be highly effective for targeting pelagic species like tuna and wahoo.

4.2 Bottom Fishing Rigs

Bottom fishing rigs are designed to present bait or lures near the ocean floor, where many species of fish feed. Common bottom fishing rigs include the Carolina rig, the knocker rig, and the high-low rig, each of which is tailored to specific bottom conditions and fishing objectives.

4.3 Live Bait Rigs

Fishing with live bait requires specialized rigs designed to keep baitfish alive and swimming naturally while minimizing the risk of tangling or fouling. Techniques such as kite fishing and live chumming are popular among anglers targeting large, offshore predators like sailfish and marlin.

5 Designing Rigs

Finding the ideal time and location to fish is only as crucial as using or creating the perfect rig, since improper presentation of the bait can prevent the fish from noticing it or from becoming persuaded to take it. The performance of rigs should be optimized for the conditions in which they are employed. Knowing which rigs are ideal for the type of setting and weather where a fisherman plans to fish is crucial. Either the rigs are clipped down (hooks are fastened behind an impact screen or bait clip) or flapping (snoods and hooks are free to flap about during the cast). When fishing close in, such as when casting from a pier into deep water or when searching for species like flounder that enter shallow water from a beach, flapping rigs are utilized. When fishermen wish to throw as far as possible, such as when fishing from a beach where longer casts are required to reach fish in deeper water, they utilize clipped down rigs. Every kind of rig has a distinct function, and some rigs go by multiple names. The categories of different rigs according to the kind of environment they are utilized in are as follows:

5.1 Surf Fishing Rigs

Fish species can be observed feeding and frolicking right off the coast of the ocean. In these situations, if one has the right gear and tactics, surf fishing can be just as successful as deep-sea fishing. Surf fishing from the beach yields some of the same fish species that may be caught offshore. Indeed, some of the best surf fishing occurs in places that aren't even boat-accessible. There are a number of factors to take into account when surf fishing, including the tide, wave height, current, and bait presentation.

5.1.1 Fish Finder Rig

Whether fishing from shore, moored, or drifting, this adaptable rig performs admirably. The leader is tied with one end to the hook and the other to the swivel. After that, the running line is passed through a plastic sleeve with a sinker snap to fasten a weight. With this setup, the fish won't feel any weight as the line is drawn out, allowing the sinker to remain on the bottom. When you sense a bite, release the hook to give the fish enough slack to properly swallow the bait. If the summer flatties are particularly wary of strangers, you can also free-spool a few feet of line. After waiting a few seconds, tighten up slowly and set the hook when the weight of a fish is felt at the end of the line.

5.1.2 Flapper Rig

The hook snoods that flap around when the rig is cast are what give this rig its name. One of the greatest rigs for short- to medium-range sea fishing is the flapper rig. It's also a nice setup to employ if you want to catch more fish by casting multiple hooks at once. The quantity of snoods to add to the flapper rig will depend on the kind of fish species you are targeting and the location of your fishing. In open waters, fishing for smaller fish is best done with two or three snoods. When fishing in regions with structures or chasing larger fish, one or two hooked snoods work well. Keeping the snoods 12 and 16 inches apart will help ensure that they don't obstruct each other as you cast out your line.

5.1.3 Dropper Loop Rig

This is a versatile rig that works well with both live and dead bait while fishing from a boat or from the shore. With this rig, one can use a bobber to dangle the bait down into the water column, or one can also use a weight to place the bait near the bottom of the water column.

5.1.4 Pulley Rig

A pulley rig can be the ideal choice in an area where the bottom is rocky. When navigating through difficult terrain, this gear helps to retain the fish on the line. Because of the downward pressure created by the fish's weight, the lead weight slides upward and prevents the rig from snagging on the bottom. It is also effective for long-distance casting because the pulley rig maintains the bait in place and increases the aerodynamics of the rig.

5.1.5 Fireball Rig

The fireball rig is designed to keep unwanted fish species from stealing the bait when bottom fishing. This rig usually has two leads, a rig body, and a sinker. The fireball rig is classified as a type of dropper loop rig, just like the Paternoster and flapper rigs.

5.1.6 Double Drop Bottom Rig

A simple saltwater fishing setup is a double-drop bottom rig. This method, sometimes known as the two-hook bottom rig, works well for catching pan fish and groupers, among other species of fish.

5.1.7 Bucktail Rig

These are excellent setups for catching flounder. Anglers often prefer utilizing these rigs because of their sensitivity, which makes for an enhanced fishing experience. The bucktail in a bucktail rig acts as a weight and a lure by resting just below the hook. In the water column, the teaser, light fly, or soft plastic lure with a connected hook will be placed further up. The bucktail rig must be fished at a 45-degree angle in order for the hook to stay near the bottom. Anglers shake the pole every few minutes and add a little piece of squid or eel to entice the fish.

5.1.8 High Low Rig

With the high low rig, two hook baits can be used at the same time. The weight is inserted below the hooks on this specific rig, not above them. This rig has a disadvantage with the weight at the bottom, even though it allows you to present two baits at once. The fish will experience resistance right away because the weight is below the hooks, unlike with a fish-finder rig, where the fish don't feel any tug when they bite. This suggests that the hook needs to be inserted immediately. Small, squishy baits, such as clams or worms, work well as bait for this rig. These are effective for several species, including stripers, fluke, and sea bass. These rigs perform best in the early or late parts of the season.

5.1.9 In-Line Snapper Rig

The in-line snapper rig is a great option for fishing for fish that are reluctant to take bait. This rig allows the bait to freely float or swim around the waves and currents just like a genuine fish would. Fish that are cautious of bait can readily eat on it without encountering any resistance due to the way it's made.

5.1.10 Whole Mullet Rig

The whole-mullet rig is an outstanding option when pursuing predatory fish such as bluefish, albacore tuna, sharks, redfish, and striped bass. This particular configuration is designed for fish that attack their prey from below. Large predators can be caught with this rig by using baitfish, such as whole mullet. This rig has a Styrofoam float attached just in front of the hook. This keeps the bait floating in the water and shields it from bottom-dwelling crabs.

5.1.11 Carolina Rig

It is a well-liked, popular, adaptable rig that works well for fishing bait off the bottom from land or when cruising in a boat. It is often referred to as a fish detector rig or a sliding sinker bottom rig. The way the rig is presented enables the bait to hang above the sinker as it rests on the riverbed or in the surf. This design keeps the line from getting tangled up on rocks or weeds and stops the fish from sensing the weight as it goes through the sinker.

5.2 Bottom Fishing Rigs

It's one of the most commonly used fishing strategies among seasoned fishermen. Both tiny and huge fish can be targeted with it. The best part is that there are a lot of methods to practice bottom fishing with varied baits, tackle, and presentations. Although there is no end to the presentations that may be used when bottom fishing, there are a few simple fishing rigs that have consistently shown to work well. The following bottom fishing rigs can be utilized to successfully target a wide variety of bottom dwelling sport fish species, depending on our application and the fishing conditions.

5.2.1 Paternoster Rig

This rig is mainly used in saltwater fishing. It makes it possible to deliver the bait above the sinker, raising it slightly above the bottom. The rig can be tied with two swivels and two hooks to create a double paternoster, or as a single paternoster. A double paternoster rig has the benefit of allowing two distinct baits to be offered simultaneously. The entire rig can be tied with a single knot, such as the clinch knot or locked half blood. The bomb or star sinker (swivel implanted at their upper end) is typically used with this rig. This rig can be utilized in a variety of saltwater settings, however it is primarily employed for beach fishing. This rig is occasionally useful for freshwater fishing as well. Sinkers come in two varieties: fixed (with a swivel) and running (with a hole). They aid in the sinking of baited line, and hefty sinkers enable you to cast farther. They are available in an array of forms and dimensions.

5.2.2 Texas Rig

It's among the most popular and functional setups for bass fishing on the lake. Because it works so well under dense cover and can be rigged with almost any type of hook, anglers love this configuration. This presentation is common to many other soft plastic rigs, so novice anglers might gain from understanding this practical configuration. The fishing line is first threaded with a weight in the shape of a bullet, then a glass or plastic bead is added, and finally the line is fastened to an offset worm hook. The hook enters the worm's head region and emerges around 1/4 of an inch down the worm's body. The worm is then rotated so that its body is now secured on the shank after being dragged up the hook in the direction of the shank. To make the rig weedless, the hook's point is then inserted back into the worm's body (Fig. 14.1).

While some anglers believe that the bead detracts from a realistic presentation given that most worms will not click, others find that the added noise or colored bead gives them an advantage in stained or muddy water because the clicking mimics a crayfish crawling over rocks and debris. Similar to a crankbait or spinnerbait, a Texas rig can also be used as search bait if it swims through a water column, yet not as successful as search bait. In order to get the worm as close to a bass as feasible, it is typically fished by casting as close to cover as possible. Since bass often attack the bait as it sinks to the bottom, it is best if the lure enters the water with as little noise as possible and with some regulated slack in the line. This rig can be used to fish for grubs, lizards, and worms, among other soft plastics, in densely vegetated areas. The equipment may go into the water when casting or flipping. This rig can be used to fish for grubs, lizards, and worms, among other soft plastics, in densely vegetated areas. The rig is allowed to enter the water as quietly as possible when casting or flipping, and a certain amount of regulated slack is maintained in the line. To see how many fish will strike the rig on the fall, someone is keeping an eye on the line. In the event that no bite is obtained, the rig is retrieved with a very jerky, sluggish action after resting on the bottom. Acquiring proficiency in this method mostly involves learning to distinguish between the sensation of the bottom and that of a fish. To ensure that the implanted hook tip passes through the body of the soft plastic bait, set the hook by violently snapping the rod.

Fig. 14.1 Texas Rig. BRDSWRD, CC BY-SA 4.0 <https://creativecommons.org/licenses/by-sa/4.0>, via Wikimedia Commons

5.2.3 Drop Shot Rig

A genuine look is presented to interested bass by drop shot fishing. This setup's design suspends the bait above the weight, putting the lure in front of your target and off the muddy bottom at a precise depth. It only takes a few knots to cast a line into the water for anglers who want to try this well-liked fishing setup. Drop shot fishing is easy: just throw or drop the rig straight down over the boat's side, reel up when the line falls slack and tighten it with the weight on the bottom, and then gently shake the rod to make the bait wriggle. A bite can be defined as pressure applied to the line, a slight tug, or a sideways movement.

5.2.4 Knocker Rig

With the use of a knocker rig, bait can be presented to fish in a way that imitates their natural prey. When the fish we are after are positioned at or close to the bottom, near a structure that allows them to escape, this is an excellent rig to employ. Before the fish has a chance to dash into the structure and break the line, the knocker rig's design helps notify an angler that there is a fish on the line. The knocker rig's greatest feature is how adaptable it is. It is effective for both inshore and offshore fishing, and it may catch a variety of fish, such as rockfish, tarpon, snapper, tilefish, porgies, and Goliath groupers.

5.2.5 Chicken Rig

This rig, sometimes referred to as a high low or deep drop rig, floats the bait vertically in front of schooling fish inside a water column, making it an effective method for attracting them. Additionally effective for fishing close to a variety of cover and structure types, including rocks, wrecks, trash, tree stumps, and reefs, is the chicken rig. Although it's most commonly used for freshwater fishing, this rig can also be utilized for saltwater fishing. Fish like yellow-eye snapper, bass, and vermillion snapper can all be caught using the chicken rig.

5.2.6 Slip Sinker Rig

The sliding sinker rig, sometimes referred to as the slip sinker rig, is an effective method for bottom fishing for fish that are known to bite lightly. Either drift fishing along the bottom or still fishing is how it works best. This rig's design facilitates the bait's ascent to a hovering position just over the bottom. Particularly effective for capturing mildly biting species like walleyes and catfish are slip sinker rigs. Fish are able to swallow the bait without experiencing the typical tug that warns them of impending danger, leading them to spit the bait out and move away. Although the rig can be put together and altered in different ways, the fundamentals stay the same. The weight of the slip sinker rig is never fastened to the line, which is its secret. In order to prevent the fish from feeling the tug of resistance when it initially grabs the bait, this permits the line to pass freely through the weight.

5.2.7 Running Ledger Boat Ridge

It is most effective when fishing from a boat and is sometimes referred to as the running rig, ledger rig, or running ledger. The running ledger boat rig is simple to put up and is made to keep the bait in the ideal spot to draw fish by pinning it to the ocean floor. This rig is particularly effective when jerking your bait across a huge expanse of marine vegetation. The hook might fully set in before the fish feels any resistance if it is allowed to swim with the bait for an extended period of time.

5.2.8 Flatfish Rig

The flatfish rig is made especially for catching certain types of flatfish. Although it's best suited for flounder and plaice, this rig can also be used to catch smaller fish. This rig's biggest feature is that it contains three snoods, so you can use three different kinds of bait. This intensifies the fragrance, which attracts fish to the rig more. To attract the fish you are targeting, you can also choose to embellish each snood with beads or sequins. If you are fishing in an area with sandy beaches and throwing short distances, the flatfish rig is a fantastic option. The drawback is that the various hooks increase the likelihood of being hooked and make it more difficult to cast out far.

5.2.9 Pickerel Rig

When attempting to target fish that are near the bottom, this rig is utilized. To distinguish the leader line from the main line using this rig, a bright wire is attached to the main line. By adding this space, you can lessen the chance of your two leader lines tangling. Two beads that you place above and below the rig assist hold the pickerel wire in place. It can now spin around the main line thanks to this. The rotation of this wire helps your bait move more organically in tides and currents. Because the wire is malleable and lightest, the pickerel rig is ideal for fish that bite light.

5.2.10 Spreader Bar Rig

Spreader bar rigs are great if one wants something that makes our bait act like a group of fleeing baitfish. These rigs are perfect for catching tuna, mahi mahi, marlin, and other large saltwater fish from your outrigger. These come in all different shapes, sizes, and colors.

5.2.11 3-Way Swivel Rig

The 3-way rig derives its name from the main swivel used in its construction and is also known as the wolf river rig. Unlike the sliding sinker bottom rig, this rig is a fixed bottom rig that doesn't slip, allowing us to place the bait a set distance off the bottom without feeding the line upon a bite. This rig is primarily used in river systems from stationary positions, whether anchored in a boat or from the shoreline. The rod tip acts as a bite indicator. Keeping the line tight from the rod to the rig ensures a direct connection, and when a bite is detected, the rod tip will twitch. One should then drop the rod tip towards the rig to collapse it back to the fish, allowing he slack to tighten before setting the hook with an upward sweeping motion. The 3-way rig is highly versatile, suitable for fishing in shallow or deep waters with real or artificial baits, and can be made using various weights to reach different depths.

5.2.12 Split Shot Rig

If you want to catch trout, bass, walleye, or panfish, this rig is a wonderful choice. When assembled, this rig is very easy to use and exposes the bait to the target fish in a very natural way. This rig is designed with a spinning configuration because it will be casting a lot of weight. A weight that is just hefty enough to let the bait drop gradually and land at or close to the bottom should be chosen. If you're seeking walleyes with this rig, add a leach; if you're hunting bass, add a floating worm.

5.3 Saltwater Rigs

There are a variety of saltwater fishing rigs that can be used to catch a variety of saltwater fish species. The type of rig you choose will depend on a number of factors including fishing conditions, target fish species, and whether you'll be trolling the open sea by boat, bottom fishing, mid-water fishing, or surf fishing from the shore. Most important, the fishing rig you choose should be designed to present your bait, or lure, in a way that looks most natural to the fish.

5.3.1 Sabiki Rigs

Sabiki Rigs are made up of several small hooks and are intended to catch baitfish. Then, tiny bits of natural bait, foil, or feathers are used to rig these tiny hooks. Plankton is the meal that baitfish eat, hence a Sabiki Rig should resemble this microscopic food and fit readily into their jaws. Sabiki Rigs are available at most tackle stores and often come pre-made with 4 to 10 hooks. While they are not very expensive, they can be challenging to store without becoming tangled. To prevent the rig from coming loose while there are fish on the end of the line, it is crucial to fasten all of the terminal tackle firmly.

5.3.2 Popping Cork Rigs

A popping cork is a type of terminal tackle used in fishing rigs that take use of a fish's excellent sense of hearing. Most tackle stores sell popping corks, which are composed of a short length of stiff wire that is threaded through a foam or cork float and a few metal or plastic beads. These are among the top redfish fishing rigs, and they also work well as trout rigs for speckled trout. Where shrimp or baitfish are active on the surface is where this setup works best.

5.3.3 Dropper Loop Rigs

Perfect for still fishing is the dropper loop rig. With this rig, you can use a bobber to dangle the bait down into the water column, or you can use a weight to position the bait near the bottom of the water column. This is a very adaptable rig that is useful for both shoreline and boat fishing with live or dead bait.

5.3.4 Pulley Rigs

A pulley rig can be the ideal choice if you're fishing in an area where the bottom is rocky. When navigating through difficult terrain, this gear helps to retain the fish on the line. Because of the downward pressure created by the fish's weight, the lead weight slides upward and prevents the rig from snagging on the bottom. Because the pulley rig maintains the bait in place and increases the rig's aerodynamics, it is also effective for long-distance casting. This rig is used to target large fish that require long casts in order to catch, particularly when reeling them in over uneven ground.

5.3.5 Whole Mullet Rig

When targeting predatory fish like bluefish, albacore tuna, sharks, redfish, and striped bass, the whole-mullet rig is an excellent choice. This setup is made especially for fish that strike their prey from the underside. If you want to catch big predators with this approach, consider adding baitfish like entire mullet. A Styrofoam float is affixed to this rig right before the hook. This keeps the bait suspended in the water, protecting it from crabs that live on the bottom.

5.3.6 Balloon Fishing Rig

The balloon fishing rig is a suitable option when fishing in conditions when the baits must be floated out across wide regions of water. It is made to catch more wind and cover a lot more ground, enabling it to travel swiftly over great distances. For fishermen who wish to experience the sensation of drift fishing but lack access to a boat, the balloon fishing setup is an excellent choice. This rig may be made in a lot of different ways, but the essential thing is that the balloon needs to be connected to the line in order for the bait to move swiftly and effortlessly.

5.3.7 Slip Bobber Rig

It's sometimes referred to as the slip float rig and makes adjusting depth simple because it can move the bobber stop rapidly. In order to hook the fish in time, it is also an extremely sensitive rig that allows one to feel the bite immediately. You can fish in freshwater and saltwater with this rig. When aiming for catfish, panfish, mackerel, snook, tarpon, trout, and redfish, it performs really well. The rig is simple to use, has a long casting range, is highly sensitive to fish attacks, functions well in both shallow and deep water, and is easily adjustable to varying depths.

5.4 Freshwater Fishing Rigs

5.4.1 Ned Rig

A straightforward and adaptable fishing rig for catching panfish, black bass, and crappie is the Ned rig. However, it works especially well with largemouth bass. Because of its highly natural appearance and motion, the Ned rig works well in high pressure locations as well as in difficult fishing situations. This is a more subtle fishing method that combines a mushroom-shaped jighead with a soft, 2–3 inch plastic lure (such as worms, crawdads, or other critters). Half of a soft plastic Senko was initially used to make the Ned rig, which was then threaded onto a mushroom jighead. A wide range of soft plastic baits, including Senkos, shaky worms, craws, grubs, shinners, and leeches, can be used with the Ned rig.

Despite its effectiveness, the Ned rig has the potential to snag easily. The brush guard head needs to be used in place of the mushroom head jighead in order to minimize snagging and enhance performance in cover like grass, pebbles, and brush. Many well used fishing methods can be used to fish the Ned rig. Casting a Neg rig into the water and allowing it to sink is the simplest method of fishing one. The Ned rig will come to rest due to the weight of the jighead. The free-floating sections of the rig will, however, rise above the water due to the soft plastic's buoyancy. When settled on the bottom, a Ned rig that uses a craw bait will appear like a crawfish defensively positioned just off the bottom. A Ned rig with a plastic worm will look like a real worm wiggling around on the bottom. When set up and fished correctly, the Ned rig is irresistible to bottom feeding bass and panfish.

5.4.2 Wacky Rig

This rig provides a lot of movement, consider the wacky rig. This rig is made by attaching the middle of a soft plastic worm to a hook by hooking it or using an o-ring. This setup allows both ends of the worms to float freely, which provides more movement and attracts more fish. Though some believe the wacky rig should only be used in the spring, it actually works year-round. It is particularly effective for catching bass. In fact, for some anglers, this is the lure of choice when targeting bass in tough situations at any time of the year.

5.4.3 Neko Rig

With the exception of adding a weight to one of the ends of the soft plastic bait, the neko rig and the wacky rig are similar. One end of the bait is secured to the bottom by the weight, leaving the other end free to float. Additionally, the rig gives the bait a distinct movement that draws fish as it descends through the water column. This rig reaches the bottom faster and has an intriguing zippy movement because to the added weight.

5.4.4 Shaky Head Rig

A straight-tailed worm with a circular, jighead-style hook make up this rig. The bait can sway back and forth as you reel it in through various sorts of cover thanks to the addition of the jighead. The worm appears to be alive because it moves in tandem with the jighead. A finesse worm might be a good match for this. Use a heavier jighead instead of a lighter one if there is a lot of movement in the water because of wind or current. This will maintain the rig's proper balance and allow the bait to continue dancing along the bottom. When fishing from a bank into open water or in locations where there is a lot of current, the shaky head rig is a very useful choice.

5.4.5 Drift Rig

Bottom fishing is a great use for the drift rig. This rig's design prevents tangling of the fishing line. When drift fishing for fluke (flounder), the drift rig is an excellent tool. This rig works best with natural baits like sand eel and squid, but it can also perform well with artificial flies, jigs, and lures. The type of fish being targeted will determine the size of the weight and hook used.

5.4.6 Slip Float Rig

A fishing line, hook, slip bobber, bobber stop, and weight make up the slip float rig, also called the slip bobber rig. The slip bobber's maximum line-up distance is limited by the bobber stop, which is positioned above it. The weight and hook are positioned underneath the slip bobber. Between the weight attached to the line below and the bobber stop above it, the slip bobber can slide along the fishing line. Upon casting this rig, the weight pulls the line through the slip bobber until it hits the bobber stop. Depending on where the slip bobber is positioned along the fishing line, the slip float rig can be utilized in shallow or deep water. A versatile rig for catching a variety of freshwater and saltwater species is the slip float.

5.4.7 Float Rig

One of the most widely used setups for freshwater or saltwater gamefishing is the float rig. You can use this setup to catch gamefish that you want to catch or to catch bait fish for upcoming fishing excursions.

5.4.8 Three Way Swivel Rig

One of the most adaptable rigs for fishing with real or fake baits in shallow or deep water is this one. Anglers that drift live bait or slow trolling frequently utilize it. Additionally, it functions effectively when fishing in current-driven water. It is also frequently used for drift fishing, shore casting, and trolling because to its extreme versatility.

5.4.9 Spinner Rig

Metal lures featuring metal blades are used to make spinner rigs. These blades rotate around the body of the lure as you cast out the line. The blades continue to spin when it reaches the water, capturing light and causing vibrations that draw in fish. Spinner rigs are an excellent tool for catching big game fish, including walleye, pike, bream, trout, and salmon. They can be utilized with a variety of fishing techniques. Use a wire spinner rig while muskie or pike hunting to prevent the fish from biting off the lure. Spinner rigs, also referred to as crawler harnesses, are an excellent choice for trolling a variety of fish, particularly walleye and panfish.

5.4.10 Bottom Bouncer Rig

The best setup for trolling walleye is a bottom bouncer rig, which can also be successful for shore fishing. It's essential to use a weight effective for the depth you're fishing to ensure success with this rig. Finding the ideal trolling speed is necessary to prevent snags and maintain constant contact with the bottom.

A bottom bouncer consists of a weighted L-shaped piece of heavy wire, with live bait, a crawler harness, or floating spinner rigs attached to the non-weighted arm via a snap swivel. The weight can range from 1/2 ounce to 3 ounces. This setup is tied to your rod by attaching the line to the eye at the intersection of the two wire arms of the bouncer. It is effective for trolling or drifting, as it presents the lure or bait rig above snag-filled areas, such as small rocks, logs, mud, sand flats, or open basins. The weighted wire feeler arm minimizes hang-ups while riding upright across underwater structures and detecting snags.

Using a bottom bouncer is a search technique that covers a lot of water efficiently, making it ideal for finding active feeding fish, especially walleyes that tend to relate to bottom structures. To fish a bottom bouncer, the angler casts or releases the line, allowing the rig to sink to the bottom. Engage the reel to tighten the line to approximately a 45-degree angle from the boat. This angle ensures frequent bottom contact, control, and feel. Avoid having the line vertical, as the 45-degree angle provides enough distance for the bouncer to work correctly and prevents spooking the fish. When you feel a bite, indicated by a subtle tap, twitch, or pressure on the line, lean the rod tip forward slightly and use a sweeping overhead motion away from the fish to set the hook. This helps eliminate any slack line and ensures a good hookset.

5.4.11 Double Jig Rig

The double jig rig is a good option when targeting crappie. Other fish including perch, bluegill, rock bass, and other panfish can also be caught with this rig. This setup is typically used by fishermen for sea fishing. For summertime fishing, this rig is most effective when targeting fish that congregate in deeper areas in schools. One can fish at two different depths and present different colored lures by employing a rig with two jigs.

5.4.12 Castable Umbrella Rig

For bass, the castable umbrella rig is particularly effective. Thus, tournament organizers have outlawed rigging. The type of umbrella rig—bladed or non-bladed—depends on the fishing circumstances. In clear, calm water, it is often recommended to utilize the non-bladed form of the lure. When there's a strong wind or muddy water, it's ideal to use the bladed version. The umbrella rig can draw bass far and wide because of its size and ability to resemble a school of fish.

5.5 River Fishing Rigs

For both novice and expert fishermen, river fishing is one of the most popular freshwater fishing pursuits. It is a low-tech activity that may be done from the shore, in a kayak, or in a canoe. There are a few most effective methods and lures for river fishing. Among the fundamental methods of river fishing are fly casting, river jigging, bottom bouncing, and upstream casting. The season, weather, water quality, and of course the game fish you are targeting will all influence the bait you choose for river fishing. However, the presentation of your bait is just as crucial as the selection of bait. A variety of fishing rigs are perfect for displaying your bait or lure in a way that will draw in river-dwelling game fish species.

5.5.1 Fly Fishing Split Shot Rig

In fly fishing, the area in the water column where the fish are hitting is referred to as the "strike zone." The majority of fish in fly fishing will only strike flies that are within this strike zone. The quality of the water, the type of food the fish are consuming, the weather, and other variables all have an impact on the strike zone. When an angler needs to add weight to their fly rig, a split shot comes in helpful. This gives an angler the chance to capture more fish by getting the fly into the striking zone faster and allowing it to stay there for a longer period of time throughout the drift.

5.6 Deep Sea Fishing Rigs

Recreational fishing from a boat at a depth of 100 feet or more offshore is known as deep-sea fishing. Specialized gear, presentation, and technique are needed for deep sea fishing. Trolling, bottom fishing, chumming, popping, and jigging with lures are the most popular deep-sea fishing methods. A range of artificial and fresh baits, such as jigs, spoons, cut bait, live fish, and squid, can be used for deep-sea fishing. Catching the majority of deep-sea game fish species requires using the right bait and fishing method. Another essential component of a good deep-sea fishing trip is having the proper bait presentation. The target game fish and the fishing method should be your main deciding factors when selecting a deep-sea fishing rig. The location and water conditions should also be taken into account while choosing a rig.

5.6.1 Two Hook Bottom Fishing Rig

The two-hook or double hook bottom rig is the best choice for saltwater anglers who intend to fish on the bottom. Although it can take some time to set up, this setup is excellent for bringing in large groupers and snappers. This rig is established by the leads that exit the 3-way swivel. Larger fish find live bait more tempting since it can swim around thanks to the two-hook rig. This rig is often more effective than the double hook drop shot rig when used with live bait.

5.6.2 Wishbone Rig

The wishbone rig is a popular tool used by fishermen to catch flatfish. This rig is particularly effective for fishing plaice, flounder, and dabs. With the wishbone rig, fishermen can utilize two hooks on the seabed simultaneously and in close proximity to one another. The benefit of this is that it helps focus the bait scent trail into a more condensed area. While larger fish may eat both at once, smaller fish will bite on each hook separately. When fishing in a tide run, this can be a very productive setup. Rough terrain is not a good fit for the wishbone rig because of its complexity. It functions best on sandy beaches because it won't snag on any undersea vegetation (Table 14.1).

Table 14.1 Optimal seasons and conditions for different fishing rigs

Name of the rig	Best Time to Use
Texas rig	All year round; especially effective in weedy or heavy cover situations
Carolina rig	Early summer and fall; best for covering large areas of water
Wacky rig	Spring and early summer; excellent in clear water and heavy cover
Drop shot rig	All year round; exceptional during post spawn and when bass are suspended
Neko rig	All year round; outstanding during post spawn and in clear water
Tokyo rig	Spring to fall; ideal in heavy cover and when punching through vegetation
Chicken rig	Less conventional, but can be effective all year round; particularly good for skittish bass
Split-shot rig	All year round; particularly good in calm water and for pressured or spooky fish
Fluke rig	All year round; particularly effective during the post spawn period

6 Conclusion

Rigging is both an art and a science, requiring a blend of technical expertise, creativity, and practical experience. By mastering the essential techniques and understanding the nuances of different rigging setups, you can greatly enhance your effectiveness as a sport angler and increase your chances of success on the water. Whether you're a seasoned veteran or a novice angler, the pursuit of the perfect rig is an ongoing journey, one that promises endless opportunities for discovery, experimentation, and, ultimately, the thrill of the catch.

15 Knots

Abstract

This chapter provides a comprehensive exploration of essential fishing knots, crucial for securing hooks, lures, and lines in angling. It examines various knots such as the Improved Clinch, Palomar, and specialized options like the Bimini Twist, detailing their strengths, applications, and tying techniques. Factors influencing knot selection, including line type, fishing technique, and target species, are discussed, emphasizing the importance of knot proficiency for angler success. Practical insights into knot tying methods, such as lubrication and trimming, ensure optimal performance under varying pressures. Whether in freshwater or saltwater environments, mastering these knots empowers anglers to handle diverse challenges confidently, enhancing fishing efficiency and enjoyment.

Keywords

Fishing knots · Knot tying techniques · Angling techniques · Fishing line

1 Introduction

A fishing knot is a particular way that a fishing line is tied to hooks, lures, swivels, or to connect two lines together. Because it guarantees a safe connection between the fishing line and other components of the fishing rig, it is a crucial ability for anglers. Fishing knots are useful for many important reasons. They have to be sturdy enough to bear the pressure of a hooked fish, keep the line from slipping or snapping, and hold up well under pressure. Different knots are used for different purposes, depending on the type of line being used, the fishing technique being used, and the unique needs of the fisherman. Tying fishing knots requires an understanding of the physics governing the distribution and transfer of forces throughout the

© The Author(s), under exclusive license to Springer Nature Singapore Pte Ltd. 2024
S. T. Mushtaq et al., *Angling in India*,
https://doi.org/10.1007/978-981-97-8876-7_15

Fig. 15.1 Improved clinch knot. Der Barbar, CC BY-SA 4.0. <https://creativecommons.org/licenses/by-sa/4.0>, via Wikimedia Commons

fishing operation. The specific arrangement of loops, twists, and wraps in each knot is designed to optimize strength and dependability. The target species, fishing conditions, and anglers individual experiences all influence the knots they prefer. Some knots are more specialized and ideal for specific tasks, like joining lines of different diameters or creating strong connections for big game fishing, whereas other knots are more versatile and frequently used.

2 Improved Clinch Knot

The clinch knot is among the easiest knots to learn. The line is passed through the eye of a hook or lure, twisted around the standing line five or six times, and then passed back through the loop that forms just above the hook eye (Fig. 15.1). A Clinch Knot could untie itself and slip, which is the problem. The Improved Clinch holds the tag end in place by passing the line back through a secondary loop that forms after the line passes through the loop at the hook eye. There is a big disclaimer though: This knot works great with monofilament or fluorocarbon line, but not with braided line. When under stress, an improved clinch knot tied with braid will unravel.

3 Palomar Knot

The Palomar knot is a popular fishing knot that is renowned for its durability and dependability. Its commonly used to attach fishing line straight to the eye of a hook, so it works well with a variety of fishing techniques and hook sizes. The Palomar knot can be used to tie monofilament or braided fishing lines (Fig. 15.2). The

Fig. 15.2 Palomar knot. Vaughan Pratt, CC BY-SA 3.0 <https://creativecommons.org/licenses/by-sa/3.0>, via Wikimedia Commons

Palomar knot, which creates a double line through the hook's eye, is well known for its strength because it evenly distributes pressure and lowers the chance of slippage. Due to its dependability and relative simplicity in tying, this knot is a favorite among many fishermen in a range of fishing environments.

4 Turle Knot

The Turle knot, so named in honour of Major William Greer Turle, an English angler from the nineteenth century who popularized the knot without claiming to have invented it, is well-known for joining a fly or hook to a leader. It can be utilized with a hook's turned-up or turned-down eye. The improved Turle knot is another name for this method of tying the knot. The line emerges perpendicular to the hook shaft, which makes this knot unique.

5 Blood Knot

The blood knot, a traditional fishing knot, is used to join two lines of the same diameter (Fig. 15.3). Anglers often use this robust and dependable knot for making leaders, tippets, or lengthening fishing lines. The blood knot forges a strong, reliable connection between the two strands. It is commonly used, for example, when attaching tippets or creating leaders, or in other circumstances where a smooth transition between two lines is desired. It is essential to practise and make sure the blood knot is executed correctly in order to increase its strength and efficacy.

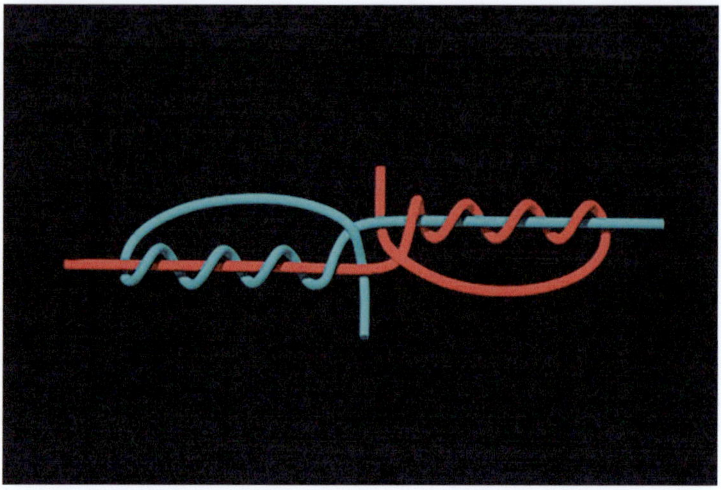

Fig. 15.3 Blood knot. Sergioscabalero, CC BY-SA 4.0 <https://creativecommons.org/licenses/by-sa/4.0>, via Wikimedia Commons

6 Surgeon's Loop

The surgeon's knot is a popular fishing knot used to join two lines, especially ones of different sizes or materials. Fishermen love it because it's an easy knot to tie that is simple, reliable, and uncomplicated. Fishermen who fish in freshwater or saltwater commonly tie the surgeon's knot. The two lines are firmly and securely joined by the surgeon's knot. Combining lines of different diameters benefits greatly from it as the additional wraps help to maintain the integrity of the knot and prevent it from slipping. Because it's fairly easy to tie, the surgeon's knot is a reliable choice for anglers looking to quickly and securely join lines together.

7 Snell Knot

One kind of fishing knot that's primarily used to attach a fishing hook to a fishing line is the snell knot (Fig. 15.4). Anglers often prefer it because it can provide a straight and secure connection, allowing for better hooksets and increased sensitivity. Applications for the Snell knot in freshwater and saltwater fishing are common. By establishing a strong and direct connection between the fishing line and the hook, Snell knots facilitate effective hooksets and lessen the chance of the hook slipping or turning. It is commonly used with hooks that have an offset or an upturned eye because it accurately aligns the hook for the best presentation and hooking potential. The snell knot is especially popular when fishing with live bait.

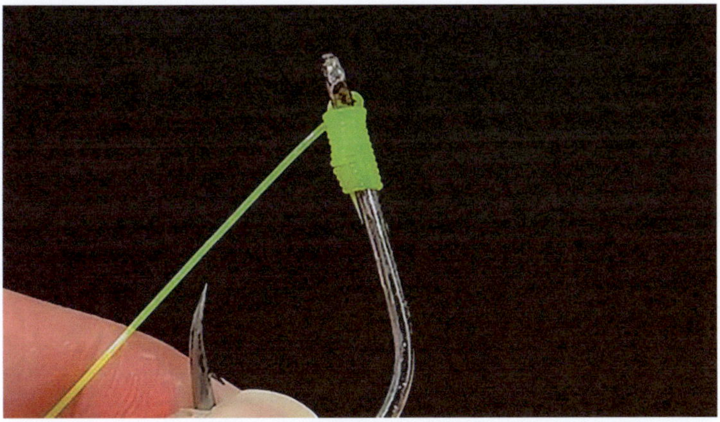

Fig. 15.4 Snell knot. Fishingknotsfast, CC BY-SA 4.0 <https://creativecommons.org/licenses/by-sa/4.0>, via Wikimedia Commons

8 Uni Knot

The Duncan knot, also known as the Uni knot, is a versatile fishing knot that can be used in a range of fishing scenarios. Anglers appreciate it for its versatility, longevity, and ease of use. The Uni knot can be used to attach the fishing line to hooks, lures, swivels, or other lines with different or equal diameters. The Uni knot is well known for its durability and dependability. It offers good knot strength and slippage resistance and solidifies the bond between the fishing line and the lure or hook. Whether two lines have the same diameter or a different one, the Uni knot is a good way to join them. Anglers like this option because it is easy to use and versatile. It is a popular choice for anglers of all skill levels because it is a flexible knot that is easy to tie quickly. Because of its numerous applications, the Uni knot is a valuable addition to any angler's toolkit. It can create reliable connections and enhance fishing excursions in general.

9 Double Uni Knot

The double uni knot is a popular and versatile knot used in fishing and angling to join two lines, usually of the same or different diameter. Connecting a mainline to a leader or joining two lengths of fishing line together is a great use for it. A double uni knot creates a strong connection between two lines. It's important to make sure the wraps sit flush and firmly against each other when you gradually tighten the knot. Because it becomes easier to tie with practice, the double uni knot is a useful knot for joining lines of similar or different sizes in a variety of angling scenarios.

10 Trilene Knot

The Trilene knot, also called the Double Fisherman's knot or the Two Turn Clinch knot, is a popular knot used in fishing and angling to fasten terminal tackle, such as hooks, swivels, or lures, to the fishing line. It is widely used due to its robustness and durability. Between the fishing line and the terminal tackle, the Trilene knot creates a stable and secure connection. The knot needs to be tightly tied in order to ensure maximum strength. With practice, the Trilene knot can be tied quickly and effectively, making it useful for fastening different types of terminal tackle to fishing line.

11 Spider Hitch

The Spider Hitch is a fishing knot that's frequently used to make a sturdy and dependable loop at the end of the line. This loop is often used to connect a leader or rig to the mainline. It is well known for its strength and ability to hold onto practically all of the original line's power. The Spider Hitch creates a strong loop at the end of the line that serves as a reliable connection point for leaders or rigs. It's important to make sure the wraps are secure and tight by gradually tightening the knot. With practice, the Spider Hitch becomes a less difficult knot to tie, making it a valuable tool for creating loops in fishing line.

12 Arbor Knot

The nail knot, a multipurpose fishing knot, is commonly used to attach a leader or tippet to a fly line. Although it is most commonly used in fly fishing, it can also be employed in other fishing scenarios where a secure and effective connection between two lines is required. The name of the nail knot originates from the tradition of using a nail as a tool, even though there are many tools available today for tying knots. Precise casting and visually appealing fly presentation are made possible by the nail knot, which forges a strong and efficient connection between the fly line and the leader or tippet. It is a sturdy knot that remains intact under stress and doesn't come undone. The nail knot is useful for joining a leader to a fly line or for changing between lines of different diameters. The traditional nail knot can only be tied with a nail or other knot-tying tool, but modern tools make the process much simpler. These tools are shaped like a small tube or cone, which facilitates wrapping the leader and passing it through the wraps.

13 San Diego Jam Knot

The SD Jam Knot, also known as the San Diego Jam Knot, is a strong and dependable fishing knot that can be used to join lines with the same or different diameters. Anglers like it because of its simplicity of use and ability to hold onto a sizable portion of the line's breaking strength. A versatile knot that can be used for both freshwater and saltwater fishing is the San Diego Jam Knot. The San Diego Jam Knot forms a strong and dependable connection between lines that have the same diameter or a different one. Its dependability and ability to hold strong knots make it suitable for a range of fishing situations. The knot's wraps increase its tensile strength and lessen the possibility of slippage by evenly distributing pressure throughout the knot. The San Diego Jam Knot is versatile and can be used to join leads to the main line, combine lines and secure hooks or lures. It is particularly useful for connecting lines of different diameters because it provides a secure and effective connection. Anglers can quickly tie the San Diego Jam Knot and be confident that it will hold firmly while fishing with a little practice.

14 Perfection Loop

The perfection loop is a multipurpose fishing knot that creates a tight, long-lasting loop at the end of a line. It is also known as the angler's loop or the loop knot (Fig. 15.5). It's commonly used to secure terminal tackle, like hooks or lures, on fishing lines. The perfection loop knot allows for maximum range of motion and provides a sturdy attachment while retaining a good portion of the line's strength.

Fig. 15.5 Perfection loop. David J. Fred, CC BY-SA 2.5 <https://creativecommons.org/licenses/by-sa/2.5>, via Wikimedia Commons

An effective loop that can be used in a range of fishing scenarios is produced by the perfection loop knot. Its sleek design enhances the natural action and presentation of lures or flies by allowing unrestricted terminal tackle movement. Additionally, the loop can be used to quickly and easily modify leads, lures, or hooks. Although it can be used for other kinds of fishing as well, fly fishing is where the perfection loop knot is most commonly used. Because it's easy to knot and retains its strength well, a lot of fishermen like to use it to create a secure loop at the end of their fishing line.

15 Dropper Loop

The dropper loop is a commonly used knot in fishing and angling, especially when setting up multiple hooks or adding extra flies or lures to the mainline. It increases the possibility of attracting fish by allowing for multiple bait presentations. The dropper loop allows you to attach extra hooks, lures, or flies in different spots by creating one or more loops along the mainline. This setup can be helpful when targeting multiple species of fish or experimenting with different bait presentations. Since the dropper loop becomes easier to tie with practice, it's a useful knot for creating versatile and flexible fishing setups.

16 Nail Knot

The nail knot, a multipurpose fishing knot, is commonly used to attach a leader or tippet to a fly line. Although it is most commonly used in fly fishing, it can also be employed in other fishing scenarios where a secure and effective connection between two lines is required. The name of the nail knot originates from the tradition of using a nail as a tool, even though there are many tools available today for tying knots. Anglers can learn how to tie a nail knot quickly and securely with enough practice. Precise casting and visually appealing fly presentation are made possible by the nail knot, which forges a strong and efficient connection between the fly line and the leader or tippet. It is a sturdy knot that remains intact under stress and doesn't come undone. The nail knot is useful for joining a leader to a fly line or for changing between lines of different diameters. The traditional nail knot can only be tied with a nail or other knot-tying tool, but modern tools make the process much simpler. These tools are shaped like a small tube or cone, which facilitates wrapping the leader and passing it through the wraps.

17 Surgeon's Knot

A popular and useful knot for tying fishing lines or leads together when angling or fishing is the surgeon's knot. It's commonly used to join two different diameter fishing lines together or to fasten a leader to the main line. Surgeon's knot used in fishing is a simpler knot to tie than the original surgical knot. This knot is well known for being sturdy and reliable. Since the second loop fortifies the knot and keeps it from slipping, it is particularly useful when connecting leaders or lines of different diameters. Anglers prefer the surgeon's knot because it's quick and easy to tie with practice, making it a popular choice for joining fishing lines or leads.

18 Albright Knot

The Albright knot, a common fishing knot, is used to join two lines of different sizes or materials. It is particularly useful for joining a lighter mainline to a heavier leader, or for adding strength or abrasion resistance to a monofilament or fluorocarbon leader. The Albright knot provides a reliable and safe connection between two lines of different compositions or diameters. Make sure the knot is securely fastened and has been tested to ensure its dependability before heading out fishing. As it becomes easier to tie with practice, the Albright knot is a helpful knot for joining lines of different sizes or materials in a range of fishing situations.

19 Bimini Twist

The Bimini Twist fishing knot is popular because of its exceptional strength and ability to create a strong loop in the standing line. Because connecting a leader or rig to the main line requires a strong loop, it is commonly used in offshore and big game fishing. The Bimini Twist knot offers a strong connection with a high percentage of the line's breaking strength. The Bimini Twist makes it possible for leads, rigs, or terminal tackle to be securely attached to the line by creating a strong, reliable loop.

Because the knot is twisted, there are no weak spots or areas where stress is concentrated because the pressure is distributed more equally. It may take more practice to become proficient at the Bimini Twist compared to other knots. To ensure that the loops are formed correctly and the twists are even, practicing the knot-tying technique is recommended. When fishing for large, strong fish, anglers prefer the Bimini Twist knot because it provides a strong connection that can withstand the intense pressure and quick bursts of energy from these fish.

20 Egg Loop Knot

When fishing or angling, a specialized knot known as the egg loop knot is used to create a loop that securely fastens bait, such as salmon eggs or other soft baits, onto a fishing hook. It is commonly used in techniques like drift fishing and bait fishing for catching salmon, steelhead, or trout. Bait is held firmly on the hook by using the Egg Loop Knot to create a sturdy and reliable loop. Carefully tying the knot ensures that the bait will remain in place during casting and retrieval. Because it becomes

Fig. 15.6 Rapala knot. Der Barbar, CC BY-SA 4.0 <https://creativecommons.org/licenses/by-sa/4.0>, via Wikimedia Commons

easier to tie with practice, the Egg Loop Knot is a useful knot for presenting bait effectively in a variety of fishing situations.

21 Rapala Knot

One kind of fishing knot designed specifically to be used with lures like Rapala lures is the Rapala knot. This is a strong, reliable knot that allows the lure to float and act like a natural in the water. Anglers who fish both freshwater and saltwater often use the Rapala knot. The Rapala knot decreases the chance that the lure will come loose while providing a strong and dependable connection between the fishing line and the lure, allowing it to move and function as intended. It comes in particularly handy for lures that have small eyelets or split rings, for which other knots might not be suitable. Ensuring that the knot is correctly tightened and the wraps are tight is essential to improving the reliability and strength of the Rapala knot (Fig. 15.6). Before using the Rapala knot in actual fishing conditions, practice it to make sure you are confident and skilled in doing so.

22 Conclusion

Because they connect the fishing line, terminal tackle, and bait, fishing knots are essential to fishing and angling. Choosing the right knot for the job is essential to strength, reliability, and success in a variety of fishing situations. Anglers can choose from a wide range of options to suit their specific needs, from specialized knots like the egg loop and dropper loop knot to versatile and frequently used knots like the Palomar and improved clinch knots. An effective fishing knot requires practice, proper technique, and meticulous attention to detail. Knowing the benefits and drawbacks of each knot is crucial when considering factors like the type, diameter, and terminal tackle being utilized. Knowing the benefits and drawbacks of each knot is crucial when considering factors like the type, diameter, and terminal tackle being utilized. Furthermore, proper knot lubrication, tag end trimming, and snug, secure wraps all enhance knot performance. Being proficient with a number of fishing knots allows anglers to confidently take up different fishing techniques and target a variety of fish types. Robust knots give fishermen the tranquilly they require to focus on the excitement and fun of fishing while still offering the necessary holding power to successfully land fish. Just like learning any other skill, mastering fishing knot tying takes practice. Consistent practice, knot testing, and familiarization with knot strengths and applications can all lead to improvements in knot-tying skills and an increase in overall angling success. To sum up, one of the most important aspects of the art of fishing is learning fishing knots. An angler's ability to successfully and confidently pursue their fishing goals is greatly enhanced by knowing and using the proper knots, whether it's for tying a dependable connection between the line and hook, creating loops for bait presentation, or fastening terminal tackle.

Live and Artificial Baits

16

Abstract

In India, live and artificial baits are integral parts of sport fishing, providing anglers with a variety of ways to lure in and capture different types of game fish. Freshwater fishing for species like Mahseer and Trout typically uses live baits, such as worms, minnows, and insects. These baits are frequently offered utilizing fly fishing, spin fishing, and bait fishing tactics, all of which call for particular rigging and presentation techniques in order to enhance performance. Popular artificial lures for freshwater and saltwater fishing include jigs, spinners, and flies. By imitating the actions and appearance of real prey, these lures can entice fish to attack. Particularly in fly fishing, artificial flies which are made to resemble insects, crabs, and other aquatic life are indispensable. Artificial lures such as spoons, soft plastics, and spinners are frequently used in baitcasting and spinning setups to attract larger fish. The species of the target, the state of the water, and personal preference all play a role in the bait or lure selection. To maximize the odds of a successful capture, it is important to understand the behavior and feeding patterns of the target fish in addition to the appropriate presentation and retrieval tactics whether using live or artificial baits.

Keywords

Bait · Artificial · Live · Fly fishing · Lures

1 Introduction

Any material used to draw fish so they can be caught is referred to as fishing bait, sometimes called fishing lure. It is typically inserted into a fish trap or at the end of a fishing hook. There are three ways that fish can identify a bait: movement, sound, and scent. Even in total darkness, a fish can detect movement in the form of a fishing

© The Author(s), under exclusive license to Springer Nature Singapore Pte Ltd. 2024

S. T. Mushtaq et al., *Angling in India*,
https://doi.org/10.1007/978-981-97-8876-7_16

lure. The fish are more receptive to and easily identify natural baits. The different types of baits that anglers use are primarily determined by the kind of fish they hope to catch, the quantity of baits available in the area, and the local regulations that establish which baits are appropriate to use where. Baits are divided into two categories, viz., live and artificial baits.

2 Live (Natural) Baits

These tiny, living fish or worms are meant to lure in their prey. While still living, these are fastened to the hook. Since live bait is a natural food source and more effective than imitations, using it to catch fish is one of the best fishing strategies. These can be bought from an aquarium store or captured with nets and traps. They are then preserved in a live well until needed (Fig. 16.1).

Fig. 16.1 *Gammarus pulex*. Picture credit: Ubaid Wani

2.1 Types of Live Bait

2.1.1 Sea Fish Baits

Sea fish baits are readily available in frozen form from tackle shops or in fresh form from supermarkets or fishmongers, and they are frequently an angler's first choice. Common and quickly reproducing species that are easy to catch and always available are usually used as bait. Sea baits come in a variety of forms, such as anchovies, sardines, herring, smelt, and mackerel.

2.1.1.1 Mackerel

Because it has a hard skin and can withstand repeated casting, mackerel is one of the most useful and effective sea fishing baits. Because it releases an alluring scent quickly, the oily flesh of the fish is also attractive to fish. It is positioned on the hook, and the line is fastened to a bobber to allow it to be fished or weighed for bottom fishing. It can be eaten whole or in the shape of tiny fillets, chunks, or strips. You can use these right away or store them in the refrigerator for later use. It is employed as bait for cod, rays, dab, turbot, dogfish, and whiting.

2.1.1.2 Sardines

Sardines are streamlined fish that are best for angling; they are usually 10–15 cm long and contain a lot of oil. It's a soft bait that needs to be hooked when frozen because it breaks easily after being cast repeatedly. Sardines are often cured in brine to harden their soft flesh. This bait will draw in target species that are farther away from the hook. When the bait is thrown into the water, it also loses scales, which frequently results in a shower of shiny scales that may attract a fish to the bait. These are effective when used as the main bait on a hook or in conjunction with an artificial lure to draw in fish. These make great bait for pike, striped bass, salmon, big catfish, and other species.

2.1.1.3 Smelt

Because they are sensitive to light, smelt are a popular little fish that are best caught at night. They smell like cucumbers, which makes them unique, and they are very fertile. These make excellent bait for capturing cod, mackerel, and pike.

2.1.1.4 Sprat

These are tiny, soft fish that resemble sardines. They are particularly useful as float fish, or when suspended beneath a float, for all species that inhabit and feed in mid-water. Because of their small size, these are typically used as bait in whole fish form. They are used to catch pollock, cod, bass, whiting, mackerel, and are reasonably priced and accessible.

2.1.1.5 Anchovy

These are forage fish, small and oily with a strong smell, that are used as bait to draw in larger prey. Because bass, bluefish, and river catfish are drawn to strong scents and feed off the bottom, this is a great option for catching these fish. They are also utilized frozen, which prolongs their time on the hook.

2.1.1.6 Crabs

Soft shelled crabs, also known as peeler crabs (crabs with a fresh soft shell that have shed their shells), are a favorite food of many fish species and make great sea angling baits. The bulk of them begin to peel in the spring, when they are most frequently found. Fish from a broad region are drawn to the scent trail that crabs release. They can be used whole or divided in half to make two baits. Crab is a common bait for catching fish like bass, cod, large rays, flounders, and eels.

2.1.1.7 Shrimp

It is among the greatest all-natural lures for fishing in saltwater. It can be utilized frozen, fresh, alive, or dead. These naturally occur in many fish's diets and can be used as bait just as well as anchovies, sardines, and mackerel. The following fish are lured with prawns: tarpon, snook, flounder, trout, pompano, and grouper.

2.1.2 Freshwater Fish Baits

The bait used for freshwater and saltwater fishing is not the same. Natural freshwater fishing baits that are most frequently used are lampreys, worms, leeches, insects, and minnows. The most popular baits used in freshwater fishing are described below.

2.1.2.1 Lampreys

Because their bodies are packed with blood, migratory lamprey, which resemble slim, long eels, make great bait. They have long been used as bait for freshwater fishing. When they defrost, they ooze blood, which draws a lot of fish, particularly pike. Lampreys have tough skins, which allow them to stay on the hook and be cast farther.

2.1.2.2 Nightcrawlers/Earthworms

Since practically all fish will eat them, these are the go-to live baits for freshwater fishing. In addition, they are easily maintained, inexpensive, and widely available. They'll live for a very long time if kept cool and slightly damp. They are typically sold as dirt or topsoil and are packaged accordingly. They can be attached directly through the hook's front, enabling them to swim and move naturally in the water. The majority of game fish consume the entire worm, regardless of its size. Earthworms are commonly threaded on a hook by anglers who use bottom fishing for catfish and other species. This method works well because it releases the worm's body fluids and odor into the water. The worms trigger an instinctive feeding response in the fish, which causes them to exhibit an erratic motion and natural odor, forcing them to bite. Earthworms or nightcrawlers are used to catch fish like pike, bass, yellow perch, trout, and catfish.

2.1.2.3 Minnows

Freshwater anglers worldwide use live minnows as an excellent bait, particularly for species such as bass, bluegill, trout, and crappies. Minnows are naturally attractive to predatory fish because they are a common food source for them. To fish in surface and deep waters, respectively, these are fastened to a hook at the end of a fishing line with a bobber or a sinker. Anglers use traps, seines, cast nets, and dip nets to catch these, which are sold in most bait and tackle stores. The majority are raised commercially and are generally simple to maintain in an aerated bucket or live well.

2.1.2.4 Crayfish

Though they can be used for most freshwater game fish species, including bluegill, walleye, carp, and catfish, crayfish are mainly used for bass. Compared to worms and minnows, they are thought to be more popular. They make excellent bait in streams and rivers and are reasonably easy to capture in traps. Usually, not much weight is used when fishing for them—just a split shot or two, if necessary. Place them in cool water with some gravel and a few rocks that resemble their natural habitat to keep them alive until you need them later. However, their claws should be clipped if they are going to be used right away.

2.1.2.5 Crickets

Particularly during the warmer months, these insects make excellent bait for fish such as crappies, brook trout, rainbow trout, and perch. Although they are typically utilized in various water columns, they typically draw surface feeder fish, or fish that feed on the surface of the water column. To keep them alive, these are hooked through the back or abdomen, taking care to avoid getting the hook too close to the animal's head, feet, or wings. They are cultivated for commercial purposes and sold in stores that offer fishing bait. It has been reported that they can survive for up to 3 months in a commercial freezer.

2.1.2.6 Grasshoppers

They make great fishing bait because they are a natural food source for many small fish, such as catfish and trout. They are inexpensive to buy and available at nearly all bait shops. Additionally, when the sun is not directly overhead, they are simple to catch at dawn or dusk. Numerous species of small and medium-sized game fish consume them. The tall grass and vegetation that line the banks of creeks, streams, rivers, swamps, ponds, and lakes are typically home to them. A sharp hook inserted through the back behind the head is used to rig them. The best time of year to use grasshoppers as fishing bait is in the early to midsummer.

2.1.2.7 Leeches

These are very attractive in the water, so they work well as baits. These parasites can be found in lakes, marshes, and sluggish streams beneath rocks or foliage. They have a hook through the tail of the sucker. They are easily found at bait stores and keep well in the fridge for several days. They do well in a bucket filled with clean water as well. Leeches are used as bait for catching pike, walleye, trout, sunfish, and catfish.

2.1.2.8 Maggots

These are the fly larval stages; they are tiny, wriggly, highly protein-rich, and simple to hook. They can be used all year round. Anglers are familiar with three different types of maggots, which vary in size: squatt, pinkie, and maggot, which are the larvae of the housefly, green fly, and European bluebottle fly. These can be kept in a refrigerator for up to 2 weeks and are sold in tackle stores.

2.1.2.9 Grubs

These are beetle larvae, which resemble hairless caterpillars. These are effective in capturing any variety of fish that forages for worms and insects to eat. Their bodies are hooked with these. When casting grubs as bait, bobbers, fly lines, and Carolina rigs are effective methods to employ. These are primarily utilized in nations where ice fishing is popular. These are effective trout bait.

2.2 Advantages

Live bait appeals to a wide variety of fish species, as it is an essential component of many fish species' food chains, and fish are more drawn to it. This makes live bait extremely effective. If one catches it themselves, live bait is usually free and less expensive than lures. The release of natural scents from live bait draws fish in from a great distance. If one doesn't intend to fish again, leftover bait can be frozen or returned to the wild. Learning the fundamentals of bait research helps one understand which prey is preferred by different fish.

2.3 Disadvantages

Fishing with live bait is messier and smellier than fishing with lures especially if one is catching it by hand. It necessitates more frequent trips to the bait shop and careful planning to ensure one has enough. It requires special storage considerations like refrigeration or circulating water. Overharvesting fish prey is a result of fishing with live bait. All kinds of fish, even small ones and ones that weren't supposed to be targeted, are drawn to it. It increases the number of deeply hooked fish, which is detrimental because it lowers the fish's survival rate, particularly when catch and release fishing. It might not always be available or might be challenging in certain fishing spots. The shelf life is short.

3 Artificial Bait

These are any artificial bait, whether made of hair, feathers, cork, wood, rubber, metal, plastic, or combinations of these materials, that are used to mimic or replace natural bait in the hopes of luring fish. Though artificial baits come in hundreds of

Fig. 16.2 Wobbler. George Chernilevsky, Public domain, via Wikimedia Commons

varieties, most can be divided into seven basic categories: spoons, jigs, poppers, crank baits, plugs, spinners, and flies.

3.1 Plugs

Plugs, also referred to as crankbait, wobbler, minnow, shallow diver, and deep diver, are composed of hard plastic (Fig. 16.2). The majority of plugs mimic baitfish, but some also mimic other fish prey such as crayfish, insects, frogs, and snakes. They flash and move, attracting fish by making noises that game fish find appealing. They come in a variety of hues and finishes, including holographic, photo, foil, chrome, prism, glass, green, and blue. Their purpose is to be constantly thrown and caught for fast, powerful blows. Plugs can hover, sink, dive, or float depending on how they are designed. The majority of plugs suspend or float on the water's surface, but when they are retrieved, they plunge dramatically. These can be divided into two categories.

3.1.1 Surface/Top Water Plugs

These are most effective in the morning and evening when fish are in shallow, calm water that is over 60 °C. The best plugs are black ones. Others, like orange and yellow for a perch, green and white for frogs, and white, blue, and chrome for shade, are employed. They are further divided into:

3.1.1.1 Crawlers

These are retrieved slowly and steadily in calm waters. They have wings installed on the side that allow them to swim over water, and they make a plopping or gurgling sound.

3.1.1.2 Chuggers

These are primarily used by bass fishermen, and they make a popping sound when they are jerked across the water's surface.

3.1.1.3 Globe (or Centre Rotating Blade)

This type of blade is primarily used by muskie anglers. Upon retrieval, the centre blade rotates, creating a bubble trail.

3.1.1.4 Rotating Tail

Anglers who fish for muskies and pike frequently use it in calm waters. The attached blade makes a plopping noise as the tail rotates.

3.1.1.5 Surface Wobbler

This device produces a clicking sound when the lure is retrieved, and it functions best in calm water.

3.1.1.6 Propeller

Also referred to as topper bait or prop bait, this adaptable lure has props on both the tail and the nose. It can be buzzed across the water to elicit an aggressive feeding response from fish, popped using a stop-and-go technique, or run slowly with a straight retrieve.

3.1.1.7 Stickbait

Also referred to as a torpedo lure, it wobbles and lacks lips or propellers. It's employed to produce the lateral movement recognized as "walk the dog" on the surface.

3.1.1.8 Flaptail

It moves back and forth on rocks and in calm water with a brass blade attached to its tail.

3.1.2 Subsurface Plugs

These come in the following categories:

3.1.2.1 Diving Lures

These are based on the body design of the lure (cupped, pointed head, flattened curve forehead) and come with attached plastic metal lips. Any kind of fish can be caught with these in any kind of water. These run between one and twenty feet or deeper.

3.1.2.2 Sinking Lures

They fall into three categories: jerkbaits, twitch baits, and gliders. Each lure has an action that is produced by a sequence of pulls, forceful jerks, or quick stop and go maneuvers. These are further divided into three categories: mid-range (5–10 feet), shallow (1–3 feet), and deep (10 feet and beyond).

3.1.2.3 Floating Minnow Crankbait

The most well-liked and adaptable crankbait that mimics a thin-bodied bait fish is the floating minnow.

3.1.2.4 Floating Shad/Perch Crankbait

This bait resembles a floating minnow and has a diving lip. It is meant to mimic the foraging behavior of larger, fatter shad or perch. Balsa wood and hollow plastic, either with or without internal shot-filled chambers to create a loud rattle, are used to make these.

3.1.2.5 Floating Lipless Crankbait

As it moves through the water, it is intended to dive based on a flattened or cupped forehead that wobbles from side to side.

3.1.2.6 Vibrating Crankbait

These have a slender body and no diving lip. They move tightly to and fro because they are fastened to the line with the eye on top of the head. These also have shot-filled internal chambers that rattle loudly.

3.1.2.7 Sinking Crankbait

Countdown, another name for sinking crankbait, is weighed so that it sinks horizontally. After the lure is thrown, it is retrieved after a countdown that starts at one foot per second and goes down to a predetermined depth. To keep the depth constant until the end of the retrieve, it employs a mid-range lip.

3.1.2.8 Neutrally Buoyant Crankbait

Also referred to as a jerk bait or suspending lure. In order to achieve neutral buoyancy, an internal weight system is used in its design. It is thrown as a random jerk-pause retrieve. When the jerk is stopped, the line in the water stays suspended and motionless.

3.1.2.9 Floating Trolling Plugs

Because they are lightweight to cast, these are mostly used for trolling. At rest, the majority of them float, and when they dive, their flattened foreheads cause them to wobble erratically and widely through the water.

3.1.2.10 Floating Jerkbait

These big, elongated plugs are designed to catch pike and muskies. They float when at rest and dive when you give them a hard tug or jerk.

3.1.2.11 Sinking Gliders and Twitch

These sink. When retrieving the twitch bait, the rod is quickly tapped or twitched, giving the lure an erratic up-and-down and side-to-side motion. The glider is retrieved with timed, brief pulls that cause the lure to dart and flash as it glides side to side or underwater walks the dog. These days, luminous plugs are a novel invention that work well in murky waters. These can be charged by shining a torch at them or by leaving them in the sun, which will cause them to reflect light that draws in predatory fish.

3.1.3 Poppers

These particular lures are designed to entice fish by skipping across the water. Their front faces are cupped, which is intended to release a stream of water when they are retrieved (Fig. 16.3). The fish are drawn in by the popping air in the cavity during their retrieval. To draw fish, the ensuing sound and splash imitate the motions of prey on the surface. The primary objective of employing a popper is to imitate any agitated organism that could be stirring or exerting itself on the water's surface. Fly fishing and spin fishing are the main uses for these. These are composed of hard plastic or wood. The two primary varieties are spitters, which have a distinct cavity and flick water when they are retrieved, and chuggers pop. The Fish Inc. Fly Half Surface Popper, Fish Inc. Scrum Half Surface Popper, Roosta Popper, Maxcatch LC22 Surface Popper, and Foam Mice Bodies are a few examples.

Fig. 16.3 Poppers. Stalane at English Wikipedia, CC BY-SA 3.0 <https://creativecommons.org/licenses/by-sa/3.0>, via Wikimedia Commons

3.1.4 Spoons

These are metal lures with an oblong shape that curves and concave on one side, giving them a shimmer and wobbling effect as they pass through the water. These are great for both casting and trolling. Julio Buel, a fisherman, is said to have invented spoon lures that were produced on a commercial scale in the 1830s. He established his own business in the 1840s and began selling spoon lures across the United States. For pike, bass, muskies, walleye, salmon, and trout, these are the most effective. Spoon fishing setup is determined by the kind of fish to be targeted, and there are different kinds of spoons. The following is a discussion of the different types:

3.1.4.1 Casting Spoons

It is the basic technique of casting in which spoons are cast out and then reeled back.

3.1.4.2 Trolling Spoons

These are thinner and lighter than casting spoons which are tied onto a rig with a diving crankbait and trolled on a long line for bottom species.

3.1.4.3 Topwater/Surface Spoons

These are cast over the cover and retrieved fast so that the lure remains on the surface.

3.1.4.4 Weedless Spoons

These are equipped with wire guards which prevent snags and can accommodate many retrieving methods including twitching and pausing. These are used in thick covers where fishes hide.

3.1.4.5 Jigging Spoons

These are used for fishes which dwell deep in the water. When it hits the bottom, the slack line is taken up until the rod tip is a foot above the water followed by short jerks up and down.

3.1.5 Jigs

Jigs are among the oldest and the most popular known to anglers. These have a weighted head on one side and a hook on the other. They sink easily making them great for bottom feeders. These are very versatile and intended to create a jerky, vertical jumping motion to attract fish. These come in all shapes, sizes, and colors. They are classified as follows:

3.1.5.1 Tail Weighted Jigs

These feature more weight near the rear end. They tend to fall through the water column quickly and are mostly used in deeper waters.

3.1.5.2 Centre Weighted Jigs

These have most of their weight in the centre causing them to fall in a sideways, sliding, or fluttering motion depending on angles, length, and width of the jig. These are best suited to shallow waters. Colors like blue silver, pink silver, gold silver which imitate baitfish are mostly used.

These are also divided into various types based on the type of fishing to be done and are defined as follows:

Football Jig

It has a football shaped head which allows the bait to not tip over and is used for offshore fishing. It can be stroked up and down off the bottom.

Swim Jig

It is designed to swim and is shaped as a heavy bullet to cut through cover, weeds, wood with ease. It can be fished right at the surface, scraping along the bottom or in between.

Flipping Jig

It is the most versatile jig which is used to flip into dense cover. It is used for making short flips, pitches, and roll casts to shallow wood, shoreline grass, and deeper grass beds.

Punch Jig

It is a specialized jig designed to punch through thick matted grass especially in backwaters where logs and debris are found.

Finesse Jig

It is the easiest jig to be used which requires the angler to be slower and more methodical. It can be target fishes, dragged, slipped under cover or bombed over structure. It is usually made of tungsten.

Casting Jig

It is a versatile jig allowing it to be used at many places. It is a basic all-purpose jig which can be skipped, pitched, and flipped. It has a head design that allows it to slide down rocking slopes without getting snaggered. Most have a flat bottom while others have a pointed one which resembles arrowheads.

Vibrating Jig

It is also known as chatter bait and can be used to fish different ways. It has a feverish pulsation which creates a visual effect to help fish get attracted to the strike. It is most effective as black and blue colors.

3.1.6 Spinners

These are among the greatest kinds of lures for luring salmon and trout in rivers and stillwaters. They were first created mainly for casting in the 1800s, and later on,

Fig. 16.4 Devon spinner. BentaxGermany, CC BY 3.0 <https://creativecommons.org/licenses/by/3.0>, via Wikimedia Commons

larger-scale commercial production of them started in the USA and Britain. They are made of a metal body, a metal spinner blade, and a wire body with a hook at one end and a line tied to the other (Fig. 16.4). The spinner blade rotates around the line as it is retrieved through the water, creating flash and vibrations that draw fish in. To draw fish, the line body and blade are painted in a variety of hues and designs. For catch and release fishing, these are ideal. On stillwaters and rivers, small spinners (sizes 0–2) are ideal for trout fishing. For trout or salmon, size 3–4 spinners work well, especially when the wind is blowing. Size 4 or 6 treble hooks are standard on most spinners, but many anglers prefer to flatten the hook or bars, or swap it out for a classic J hook, which makes it easier to release the hooked fish and put it back in the water. These are available in several varieties, as listed below:

3.1.6.1 In-Line Spinner

These feature a blade or multiple blades which spin around a central wire or axis of the bait.

3.1.6.2 Tail Spinner

These versatile spinners can be fished horizontally or jigged vertically for large and small species respectively. They feature a weight-forward design where the body or weight of the lire, the line tie, and a treble hook are at the front of the bait and spinner blade extends off the back of the bait by a swivel.

3.1.6.3 Beetle Spinner

It is a miniature of the overhead arm spinner with a single blade connected to the wire via a swivel which allows the blade to rotate when pulled through water.

3.1.6.4 Live Bait Spinner Rig

This includes an extended harness with a single or multiple rotating spinner blades, colored beads and a single or multiple baited hooks. The rig is connected to a sliding sinker setup or bottom bouncer weight which makes the setup more suitable for trolling or drifting. The working of spinner blade flash and vibration with the allure of live bait makes it a unique and productive tool for targeting a variety of both cold and warm water fishes.

3.1.7 Flies

Traditionally used in both fly and spin fishing, flies are a type of fishing lure. Lures for fly fishing are made up of only one hook and one skirt. Furs, feathers, or thread are used to bind these lures so they resemble insects, crustaceans, or other prey. These work well in places where fish congregate at the water's surface to feed (Fig. 16.5). The first written record of flies and fly-fishing dates back to 200 A.D. in Elian's Natural History. The phrase "artificial fly" was originally used in *The Compleat Angler* by Izaac Walton. The fourth edition of John Denny's *The Secrets of Angling* released in 1962 is the first edition known to have artificial fly illustrations. Certain fly lures are made to float. Some fly lures are designed to float (surface

Fig. 16.5 Various flies used as bait in fly fishing: (i) Dungeon streamer fly; (ii) Cone head black wooly bugger; (iii) Epoxy ant dry fly; (iv) Grasshopper fly; (v) Kelly galloup streamer; (vi) Fly box. Picture credit: Dr. Musadiq Allaqaband

flies), while others are designed to sink (subsurface flies). On the basis of the prey of the target fish, various kinds of flies are used. These are classified as:

3.1.7.1 Dry Flies

These are the types of flies that fish on the water's surface by creating the appearance of a newly hatched fly or a terrestrial insect that has fallen into the water (Fig. 16.6). They are knotted using materials that are either buoyant or have a low water absorption capacity on light-weight hooks. Typically, they have protruding legs, feathers, and fibers to enhance their surface area and facilitate buoyancy. Most people classify these as freshwater flies. The Elk Hair Caddis, Tasmanian Red Tag, Blue Winged Olive Quill, Royal Wulff, Orange Stimulator, and Adams are a few examples.

Fig. 16.6 Trout fish flies: (i) Nymphs (Stonefly and Mayfly); (ii) Larvae and pupa (Caddis); (iii) and (iv) Dry flies (Caddis, Winged ant, stone fly, and may fly). Picture credit: Ubaid Wani

3.1.7.2 Wet Flies

These are intended to sink beneath the water's surface. These have been tied in a multitude of patterns to resemble bait fish, drowned insects, nymphs, pupa, and larvae (Fig. 16.6). Grizzly King, Professor, Partridge, Alexander, Woolly Worm, and Orange Soft Hackle are a few well-known examples.

3.1.7.3 Streamers

These mimic leeches and other aquatic life as well as small fish. They work wonders when it comes to catching bigger fish in deep pools. Woolly bugger, Mickey Finn, Black conehead, Royal Coachman, Egg-sucking leech, Muddler minnow, and Schenk's White minnow are a few examples.

3.1.7.4 Emergers

These are the flies that fish on the film that covers the surface, or that hover just above or slightly below it. They mimic a half-sunk terrestrial that has been blown onto the water, or an aquatic fly that is hatching and attempting to break through the surface film. To keep the fly floating, they typically have a hackle wound around a wing post or a flotation material as a wing, such as foam, polypropylene, etc. Because of the way it's made and the natural oils it contains, it floats by itself. These are classified as trout flies for freshwater use. Blue-winged olives, soft hackles, Barr emergers, pale morning duns, parachutes, and sparkle pupas are a few examples.

3.1.7.5 Nymphs

These are subterranean flies that are meant to depict the insect in its larval stage of development (Fig. 16.6). Since nymphs are the primary food source for fish, it's a good idea to bring nymphs when fishing for smallmouth bass or when your primary objective is to catch fish. Brook's Montana Stonefly, Biot Midge, Beas Head Prince, Gold Ribbed Hairs Ear, and Pheasant Tail are a few examples.

3.1.8 Buzz Baits

These are topwater lures which are great for fishing fast and covering water such as willow grass, reeds, flooded bushes, and standing timber. These are best known for fishing in shallow water, but can be effective in clear water as well. They are built on metal wire frame and sink to the bottom. The frame is bent into a J shape with a blade rigged on the short wire up top and a hook molded into a lead head on the bottom longer wire. The tie up is provided by an R shaped bend at the front cover. The most common size used is half ounce. The best equipment for fishing using a buzz bait is a braided fishing line. These come in certain variations such as double bladed, in line and toad hybrids.

3.1.9 Soft Plastics

These are popular elastomer based fishing tackle with a flexible and flesh like texture. These are economical costing less than plugs and spinner baits. These are versatile and can be used throughout the entire water column and in both freshwater and saltwater habitats. To make soft baits, the plastic is heated into a liquid form, then poured in a mould to replicate the shape of a grub, worm, bait fish, crawfish, lizard, frogs, and insects. Additional ingredients to appeal to the fish's senses such as scents, layered colors, metal flakes, and flavors are also incorporated while making the bait. These come in the following forms:

3.1.10 Worms

The soft plastic worm technology was introduced in 1949 by an Ohio based lure-maker who developed grubs, worms, crawfish, tubes, etc. These come in various sizes from a few inches for trout and panfish upto 12 inches for bass and pike. The main fish attracting action components of the worms are the texture (ringed, ribbed, or smooth), which affects the sink rate and the tail (ribbon, twister, paddle, or straight) which provides vibrations when the worm is moved.

3.1.11 Grubs

These are composed of soft plastic round body either ringed, ribbed, or smooth combined with single curly tail, double curly split tail, paddle or straight tail. These come in lengths from 1″ upto 12″ and many color combinations. They are most commonly topped on a jig or as a trailer or an inline spinner. They are used for fishing by a drop shot, split shot, or a Carolina rig.

3.1.12 Baitfish

These mimic forage fish and come in numerous sizes and colors. Most of them feature a paddle tail which wiggles when retrieved, while others have a curly and forked tail which gives them swing action.

3.1.13 Tubes

These are open ended rounded soft hollow plastic bodied bait with a series of tentacles on the base. The body is smooth or may have a ribbed exterior. The interior hollow design holds liquid or paste scents. Upon casting a tube, it displays a spiral action on fall with tentacles undulating and displaying an injured bait fish look. These can be rigged using a Texas, Carolina, or a drop shot rig.

3.1.14 Crawfish

These are used along rocky bottoms and are deadly on bass. These mainly feature the pincers which when tipped on a jig gives the bait a realistic defensive posture by raising its claws which sends bass a signal to feed.

3.1.15 Lizards

These are mostly used during spawning season of bass and come in a wide variety of colors and scents.

3.1.16 Leeches and Reapers

These are the basis variation of a soft plastic grub with a rounded head and body leading to a soft thin membrane on sides. Small reapers resemble a leech while the large ones mimic bait fish.

Fig. 16.7 Swimbaits. Kananlures, CC BY-SA 3.0 <https://creativecommons.org/licenses/by-sa/3.0>, via Wikimedia Commons

3.1.17 Frogs

They are weedless with hooks pointed against the soft spongy body and see natural. These are used mostly for summer while fishing for bass.

3.1.18 Magnum

These were first produced as large soft plastic lures in the 1980s by small basement companies for pike and muskies. Presently these come in hundreds of variations, colors, and combinations.

3.1.19 Swimbaits

These are a class of fishing lures designed to primarily imitate the underwater swimming motions of baitfish (Fig. 16.7). These originated in the 1980s as lures designed to imitate rainbow trout which the largemouth bass fed on. These mimic a baitfish which the predatory fish would eat. These can be classified as:

3.1.19.1 Hard Bodied

These are made of wood, resin, composite materials which can be carved, molded, and segmented to provide the hard material the ability to swim and glide. These are designed to have a swinging action which provokes a predatory instinct from the fish causing it to strike the lure. The action of the lure varies based on the number of joints in the bait. These are further classified as:

Glide Bait

A hard bodied bait with a single joint which provides a serpentine or gliding action in the water which can be fished on a variety of retrieve methods and speeds.

Multi Jointed Bait

It provides a more natural swimming appearance to the bait but can also provide some unique action with different retrieve styles and speeds.

3.1.19.2 Soft Bodied

These are made out of rubber/soft plastic molded and rigged with a weight and hook built into soft swimbaits. These are further classified as:

Paddle Tail

These are most commonly used by anglers and come in many shapes and sizes. These are used as a trailer for a spinner bait, chatterbait, or underspin. It comes in either a hollow or a solid body.

Line through

Large swim baits which allows the line to run through the body of the bait to keep the fish from using the bait as leverage to throw the hook.

Top Hook

These utilize a large, strong, jig style hook which protrudes from the top of the swim bait. It has a wedge style tail for use in cold water.

3.2 Advantages

1. Artificial bait is a one-time investment that is reusable.
2. Because they don't require special storage like soil, live wells, or water tanks, the tackle box weighs less.
3. There is no smell or dirtiness to these.
4. By reducing deep hooking, lures can increase the survival rate of fish.
5. Less bait is used by lure anglers, preventing overharvesting of bait species stocks.
6. As a result, there are more large catches and fewer undesired fish on the line.

3.3 Disadvantages

1. Choosing the right artificial baits for a given species of fish can be difficult, which can lead to a cluttered tackle box.
2. These could become entangled in rocks, trees, and weeds.
3. At night or in muddy waters, these are useless.
4. The proper weight, hook size, and color combination for lures is essential, and it can be challenging for novices with insufficient experience.
5. The cost of lures can exceed that of live bait.
6. In order to draw in fish, it is necessary to move them continuously when using them.
7. In colder waters, lures are ineffective as baits.

4 Criteria for Selecting Bait for Fishing

There are some things to be considered while selecting a fly for fishing. These include.

4.1 Fly Size

There are a variety of fly sizes, and generally speaking, the larger the hook, the lower the number, and vice versa. A size 22 hook would be regarded as small, and a size 6 as the largest. The most popular sizes range from 10 to 18.

4.2 Barbed or Barbless

To keep the fish from escaping, the majority of flies have a tiny barb located directly behind the hook point. With the use of pliers, the barb is clipped down, making it simpler to release the hook and allowing the fish to go back into the water more quickly. This also prevents the barb from hurting the fish when the hook is taken out.

4.3 Weighted or Un-Weighted

To cause flies to sink or change their swimming motion, more weight can be applied to them. This weight can be hidden beneath the dressing, like lead wire or sheet, or it can be a component of the dressing, like dumbbell eyes, a goldhead, etc. These two later forms are not only attractive to fish, but they also add weight. The dumbbell eyes resemble the eyes of a bait fish, and the goldhead is bright and shiny like many nymphs.

4.4 Hook Type

Hooks are made with various weights and shapes to best fit the kind of fly that will be tied. For example, a long shank hook will be used to tie a lure, or fish imitation, since it will mimic the long, thin shape of a fish; shrimp are tied onto curved hooks to resemble hunched shrimp, etc. To change a hook's weight and strength, different gauge wires are used during manufacturing. To help with flotation and reduce weight, a dry fly hook made of fine wire is used, and heavier hooks are used to tie flies intended for larger fish in order to increase their strength.

4.5 Quality

There are different types of flies. Avoid cheap ones that are badly tied, made of inferior materials, have flimsy hooks, and break easily. It is best to source well-tied flies on high-quality hooks made of high-quality materials.

5 Conclusion

Choosing the right fishing bait, whether live or artificial, is crucial for successful angling. Live baits offer the advantage of natural scent and movement, enticing a wide range of fish species. They can be sourced from natural habitats or purchased from stores, offering versatility in application. However, they require careful handling and storage to maintain their effectiveness. On the other hand, artificial baits mimic natural prey and come in various forms like plugs, jigs, and spinners. They offer durability and can be customized to target specific fish behaviors and environments. Artificial baits are reusable and eliminate the need for constant trips to bait shops. Ultimately, the choice between live and artificial baits often depends on fishing conditions, target species, and personal preference. Both types have their advantages and considerations, ensuring anglers have a diverse toolkit to adapt to different fishing situations. Whether casting into freshwater lakes or saltwater shores, understanding the nuances of bait selection enhances the chances of a successful fishing trip.

Handling of Fish

17

Abstract

Angling is a popular recreational activity worldwide, cherished for its relaxation and connection with nature. The practice of catch-and-release has emerged as a critical conservation strategy advocated by environmentalists and fishery managers to sustain fish populations. Proper fish handling techniques are essential to minimize stress, injury, and mortality post-catch. This chapter explores effective strategies such as minimizing fight time, using low-impact equipment, avoiding air exposure, gentle handling, managing barotrauma, and reviving fish before release. Adhering to these techniques not only enhances the survival rates of released fish but also promotes sustainable angling practices crucial for the preservation of aquatic ecosystems.

Keywords
Angling · Catch and release · Fish handling · Conservation · Sustainable practices

1 Introduction

Angling is a very satisfying and relaxing activity and a hobby for millions of people across the globe. All anglers do not catch fish with an intention to relish it but do it merely to achieve a rewarding fishing experience (Fig. 17.1). In this technique, anglers immediately release the fish, unharmed back to the water from where it is caught. This practice known as catch and release has been advocated by the environmentalists and fishery managers worldwide so that the population of fishes does not decrease significantly. This technique improves native fish populations by allowing them to grow and reproduce and also provides an opportunity for anglers to enjoy fishing. It also ensures that this recreational opportunity will last for generations to come. However, if fishes are not handled properly this practice can prove

© The Author(s), under exclusive license to Springer Nature Singapore Pte Ltd. 2024
S. T. Mushtaq et al., *Angling in India*,
https://doi.org/10.1007/978-981-97-8876-7_17

Fig. 17.1 Angler gently handling a freshly caught trout by the stream. (Picture credit: Ubaid Wani)

detrimental for fishes as they can undergo physical injury, intense stress, and in many cases, death. Releasing a fish doesn't guarantee its survival, but proper handling can reduce the stress and give it a better chance to live and reproduce. Therefore, it is critical to adhere to best practices when handling fish during fishing to reduce stress, harm, and mortality (Palme et al., 2016). Keeping the fish submerged in water while removing the hook, handling the fish with damp hands to preserve their protective slime layer, paying attention to hook setting to prevent deep hooking, and cutting the line when a fish has been deeply hooked are some of these techniques that reduce air exposure. The adoption of these optimum handling procedures by anglers is also greatly aided by outreach and education initiatives. By implementing these best handling practices, anglers can significantly improve the survival rate of released fish and contribute to the conservation and sustainability of fisheries resources. Furthermore, it is important to fish in locations with low predator densities to improve the chances of survival post-release. Overall, by following best handling practices in angling, anglers can have a positive impact on the well-being and survival of fish, contributing to the preservation of fish populations and the overall health of aquatic ecosystems. The following steps have been suggested for proper handling of fish once it is caught.

1.1 Minimizing Fight Time

Minimizing the fight time when catching fish is an important aspect of proper handling and releasing practices. This can be achieved by using appropriate tackle and techniques that enable anglers to land the fish quickly, reducing the physical exertion and stress placed on the fish. By minimizing the fight time, anglers can significantly reduce the negative impacts on the fish's health and increase their chances of survival post-release. Additionally, minimizing air exposure time and physical damage while handling the fish is also crucial. One should try to land the fish as quickly as possible as exhausted fish have a difficult time recovering when caught and are more susceptible to predation. Moreover, fighting or playing a fish to exhaustion dramatically increases the chance of dying. To minimize stress in fish, it should also be made sure that large species should be landed alongside the boat within 20 min of being hooked. However, if that is not possible, one should try using heavier tackle to reduce the amount of time it takes to catch and release the fish.

1.2 Use of Low Impact Equipment (Lures and Tackle)

The employment of low impact equipment such as lures and tackle is integral to ensuring the preservation of fish health during the angling process. Such gear is specially designed to cause minimal harm to fish. For instance, the adoption of barbless hooks is a significant step towards ethical angling. Barbless hooks are easier to remove from the fish, thereby diminishing tissue damage and improving post-release survival rates (Barthel et al., 2003). Flies and single hooked lures have been known to cause the lowest impact. Using artificial lures in place of baits have proven to reduce fish mortality. Likewise, the utilization of circle hooks can also lessen the likelihood of deep hooking. Circular hooks, in particular, are recommended when using live baits. Deep hooking is intrinsically deleterious to fish, often leading to high mortality rates. By reducing the chance of such harm, circle hooks contribute to fish conservation. The shift towards low impact equipment is indicative of the change in angling practices towards sustainable and conservation-friendly methods.

To minimize stress on the fish, large species should be landed or brought alongside the boat within 20 min of being hooked. However, if that is not possible, one should try using heavier tackle to reduce the time it takes to catch and subsequently release the fish. The use of release tools (needle-nosed pliers, hemostats, and hook removers) are essential tools for quick and efficient removal of hooks. Never rip the hook out of a fish. Instead, reach into the fish's mouth and grasp the hook shank with fingers or pliers. Lift the fish and rotate the hook shank down and shake gently, allowing the fish to slide off the hook. If the fish is hooked deeply in the gills, throat, or stomach such that it cannot be removed easily, cut the line as close to the hook as possible. The hook will rust off the fish the fish in a few months.

1.3 Hold the Fish Close to Water

Holding the fish close to or in the water whenever possible is highly beneficial in ensuring its well-being. Air exposure can be a significant stressor potentially leading to increased fish mortality (Schreer et al., 2001). Lifting or suspending the fish unnecessarily out of water can cause internal injuries, as fish are not structurally adapted to support their body weight without water's buoyancy. Moreover, keeping the fish close to water can help maintain its body temperature, reducing the risk of thermal shock, particularly in warm weather conditions. Therefore this act of keeping the fish close to the water is vital for the survival of fish during catch and release.

1.4 Avoid Air Exposure

Most of the fishes are unable to respire unless their gills are submerged in water. Keeping a fish above water for too long prevents it from breathing and further adds to the angling induced stress. Avoiding air exposure is critical for the survival of fish in the angling process. Air exposure is a significant stressor for fish, which can lead to increased mortality (Schreer et al., 2001). Even brief air exposure in exhaustively exercised fish like rainbow trout can lead to increased mortality rates (Ferguson & Tufts, 1992). Therefore, it's pivotal for anglers to minimize air exposure whenever possible to enhance survival rates of catch-and-release fish.

1.5 Handle Fish Gently

Handling of fish in angling requires meticulous attention to specific practices to ensure the minimization of stress, injury, and mortality rates. A fish that has been caught should be grasped by the tail using one hand and the other hand should be used to support the fish from underneath towards the head (Fig. 17.2). This restricts its ability to move, easing control and also minimizes the loss of mucus from unnecessary contact. Dragging a live fish across rocks and sand can cause damage to gills, eyes, and the protective mucus barrier. When handling a fish, one should remember to use wet hands or rubberized gloves. Handling a fish with a dry hand removes this protective barrier (mucus) and the fish becomes highly susceptible to waterborne diseases and parasites. It should also be kept in mind that one should never hold the

Fig. 17.2 Angler Ubaid demonstrating correct fish handling

fish by gills as this can cause serious injury to the fish. Deep hooking, an avoidable occurrence with careful hook setting, significantly contributes to post-release mortality in various species. It is an adverse incident affecting a wide range of angling species (Bartholomew & Bohnsack, 2005). Using appropriate equipment is another essential factor to be considered while handling a fish.

1.6 Managing Barotrauma

Barotrauma refers to injuries caused by increased air or water pressure, such as changes from diving or flying, and it's a common issue in catch-and-release fishing. When fished from deep waters, the sudden pressure change can cause fish to experience barotrauma, leading to bloating and inability to swim back down, impacting their chances of survival. It can also cause bulging eyes, distended organs, eversion, prolapse, torsion, and volvulus of the stomach and serious conditions such as hemorrhaging of internal organs, hematomas, and loss of vision. Fish caught in more than 30 feet of water must be brought to the surface slowly and examined for possible signs of barotrauma. Also released fish exhibiting signs of barotrauma should be lowered back to depth as soon as it is bought to the surface.

Anglers can manage barotrauma in several ways. The first is through fizzing or venting, which involves puncturing the swim bladder with a sharp, hollow object/hypodermic needle to release the trapped gases. Although this method can be effective, it requires knowledge and precision to avoid further harm to the fish. Another method is the use of descending devices such as drop weights, release weights, and baskets, which bring the fish back to depth quickly. This allows the fish to recompress naturally, alleviating the symptoms of barotrauma. Using descending devices increased the survival rates of rockfish experiencing barotrauma (Butcher et al., 2012). It is also essential to reel in and release fish as quickly as possible to minimize the impacts of barotrauma. Reducing the fight time can lessen barotrauma symptoms (Brown et al., 1994). While these methods may help with immediate decompression, they often lead to organ damage and should be avoided. These methods do not increase survival but can cause injury, stress, and even death in serious cases. Therefore, avoiding fishing in deep waters where barotrauma is more likely to occur can prevent this condition (Drum & Gitschlag, 2006).

1.7 Reviving and Releasing the Fish

Reviving and releasing the fish with care is essential to ensure its survival. Fish that undergo exhaustive exercise due to an angling event can experience severe physiological disturbances, including a disturbance in the balance of ions in their blood, acidosis, and hypoxia (Kieffer, 2000). Moreover, stress and injury from the catch can cause fish to be disoriented and weakened, making them vulnerable to predators or unable to swim away properly. If not revived properly, it can result in a dead or stressed fish who might fail to survive for long. The reviving process involves allowing the fish to recover from the fight and acclimate to its environment again before being released. The best way to revive a fish before releasing it is to cradle the fish in a swimming position and moving it gently back and forth to force freshwater through the gills. One has to make sure that its ventilating before release. This way the fresh oxygenated water can pass gently over the gills and help it to breathe normally. In fast moving waters, consider moving the fish to calmer waters where it can recover and swim away on its own. This practice has also been proven to significantly increase the chances of survival post-release.

2 Conclusion

In conclusion, maintaining aquatic life and ecosystems greatly depends on the ethical treatment of fish during fishing. Fish handling techniques play a crucial role in post catch survival of fish. One should encourage the use of such techniques/practices which reduce human impact on fish intended for release. The use of optimum handling techniques, such as avoiding deep hooking, holding the fish with wet hands, and exposing it to as little air as possible, greatly reduces stress, injury, and mortality in fish that are released. Another important factor is the application of species-specific, low-impact tools and techniques. Furthermore, knowing the unique requirements for managing various fish species facilitates the application of suitable and efficient handling methods. Fisheries sustainability is enhanced when suitable equipment is used in conjunction with knowledgeable handling techniques, as these factors have a stronger impact on fish survival after release. These initiatives can be strengthened by public education and outreach initiatives, which promote an ethical and responsible culture. The future of sport fishing/angling lies in the hands of individual anglers who should be responsible enough to abide by the regulations of the state/area where they consider fishing. By following fishing regulations they are building and conserving the fish populations for future generations of anglers.

References

Barthel, B. L., Cooke, S. J., Suski, C. D., & Philipp, D. P. (2003). Effects of landing net mesh type on injury and mortality in a freshwater recreational fishery. *Fisheries Research, 63*, 275–282.

Bartholomew, A., & Bohnsack, J. A. (2005). A review of catch-and-release angling mortality with implications for no-take reserves. *Rev Fish Biol Fisheries, 15*, 129–154. https://doi.org/10.1007/s11160-005-2175-1

Brown, I., Sumpton, W., McLennan, M., Mayer, D., Campbell, M., Kirkwood, J., et al. (1994). An assessment of the effects of commercial line fishing on the coral trout *Plectropomus leopardus* population on the great barrier reef. *Fisheries Research, 102*, 305–313.

Butcher, P. A., Broadhurst, M. K., Hall, K. C., Cullis, B. R., & Raidal, S. R. (2012). Assessing barotrauma in angled snapper Pagrus auratus: Interactions between hooking location and air exposure duration. *Journal of Fish Biology, 80*, 1085–1104.

Drum, A. S., & Gitschlag, G. R. (2006). Deep reef fish surveys by submersible on Alderdice, McGrail, and Sonnier Banks in the Northwestern Gulf of Mexico. *Gulf Research Reports, 9*, 55–69.

Ferguson, R. A., & Tufts, B. L. (1992). Physiological effects of brief air exposure in exhaustively exercised rainbow trout: Implications for "catch and release" fisheries. *Canadian Journal of Fisheries and Aquatic Sciences, 49*, 1157–1162.

Kieffer, J. D. (2000). Limits to exhaustive exercise in fish. *Comparative Biochemistry and Physiology Part A: Molecular & Integrative Physiology, 126*, 161–179. Just like humans, fish build up lactic acid on getting exhausted especially during a fight.

Palme, C. A. D., Nguyen, V. M., Gutowsky, L. F., & Cooke, S. J. (2016). Do fishing education programs effectively transfer 'catch-and-release' best practices to youth anglers yielding measurable improvements in fish condition and survival? *Knowledge and Management of Aquatic Ecosystems, 417*, 42.

Schreer, J. F., Cooke, S. J., & McKinley, R. S. (2001). Cardiac response to variable forced exercise at different temperatures: An angling simulation for small mouth bass. *Transactions of the American Fisheries Society, 134*, 653–665.

Water and Sun Safety 18

Abstract

To have a safe and enjoyable experience, anglers must be aware of water and sun safety precautions when fishing in India's different aquatic ecosystems. Both basic swimming skills and the wearing of a life jacket are required when fishing from boats or in close proximity to deep waters. Anglers should also use sunscreen, sunglasses, and a hat to shield themselves from the sun's dangerous UV rays. It's important to stay hydrated, especially when fishing in hot weather, so drink lots of water. It's also critical to be cautious of the weather and to steer clear of fishing during storms. By putting safety first, anglers can enjoy the excitement of a successful catch while lowering the risks of exposure to the sun and water.

Keywords
Swimming skills · Life jacket · Sunscreen · Shield · UV rays · Hydrated

1 Introduction

Fishing is a relaxing and fun way to pass one's leisure time and experience the beauty of nature. It becomes even more enjoyable when combined with activities like boating, hiking, wildlife viewing, and photography. It is not a dangerous sport, but as it involves venturing in water, one should be careful and follow some safety measures before going out to fish.

© The Author(s), under exclusive license to Springer Nature Singapore Pte Ltd. 2024
S. T. Mushtaq et al., *Angling in India*,
https://doi.org/10.1007/978-981-97-8876-7_18

1.1 Check the Weather Forecast

Before planning to go for angling, one should always check the weather forecast especially in high altitude areas which are prone to landslides. One should take note of other weather conditions that may impact the fishing trip—for example, reduced visibility from fog or rain, or risks to one's safety due to strong winds, thunderstorms, or lightning. In case of angling in marine waters, one should check the wave conditions by reviewing the seas and swell section of the coastal waters forecast for the coastal zone where one is planning to go fishing. Seas or swell greater than 1.5 m are large enough to be potentially hazardous. The systems that generate these waves can be far off the coast, therefore these waves can come as a surprise even if it is sunny. The weather forecast issued daily by Indian Meteorological Department provides important information to help plan outdoor activities and prevent any untoward incident due to bad weather or the overexposure to sun.

1.2 Basic Boating Safety

When fishing from a boat, one should make sure that all the required equipment and a first-aid box are handy. One should have a proper know how of using the rescue devices as well. Stay seated as much as possible while in a boat. Never overload a boat and always evenly distribute the load. Also maintain a safe distance from low water dams and other restricted areas. Travel slowly in shallow areas and areas of flooded trees.

1.3 Choose Fishing Area Wisely

It is very important to choose the fishing area wisely. Whenever possible, one should try to fish from a shady spot. When fishing from the shores of a lake or river, one should try to find a spot underneath trees or structures that will keep the sun off one's skin. If one is surfcasting, a chair and a beach umbrella should be taken where one can sit and enjoy fishing without getting exposed to sunlight. If the boat has a shaded area (cabin) one can take a few breaks from the sun and relax in the shade. One should also remember to avoid fishing areas that have been labelled as prohibited or restricted.

1.4 Choosing the Proper Time for Fishing

It is a well-established fact that the sun is at its peak/strongest points during the middle of the day (10:00 am to 4:00 pm). If one plans to go fishing during these hours (10:00 am to 4:00 pm), one is most likely to get a painful sunburn because the sun is most intense. However when the sun is at an angle (before 10:00 am and after 4:00 pm), light travels through most of the earth's atmosphere, which blocks some of the harmful UV rays. It is encouraged to try fishing either in the early morning hours or around dawn when one might also get the best catches.

1.5 Sun Protection

Sun protection is a critical but often an overlooked part of safety during angling. While some exposure to sunlight can be enjoyable, too much can be risky. If not properly protected, the sun's ultraviolet rays (UV) rays can damage a person's skin in less than 15 min. Therefore, sun safety during angling involves an integrated approach to sun protection to avoid overexposure to UV radiation which often results in painful sunburn. Sunburns significantly increase one's lifetime risk of developing skin cancer, premature aging of the skin, cataracts, eye damages, and immune system suppression. The American Academy of Dermatology recommends using a broad spectrum sunscreen, which protects against both UVA and UVB rays to be applied at least 15 min prior to going outdoors for fishing. One should try finding a sunscreen that is atleast 30 SPF, offers multi spectrum protection, and is water resistant so that the sweat or water won't cause it to run off when one is constantly exposed to splashes and sprays. It is recommended to reapply sunscreen every 2 h for longer protection. Another great way of protection from sun is using a sun shade which can be installed on the boat and is an excellent investment for anglers to enjoy fishing. It is also good at mitigating the wear and tear on the boat from regular sun exposure (Fig. 18.1).

1.6 Hydration

Dehydration onboard can be problematic as it may lead to heat exhaustion or heat stroke. Staying hydrated is believed to help reduce the damage caused by sunburns because the sun burnt skin is dried out and water helps to restore and replenish the skin cells. If one plans to carry a water cooler along, try packing sparkling water, fresh-squeezed juices, or coconut water. Even simple water infused with berries, cucumbers, or citrus can be refreshing.

1.7 Using Tackle Safely

One should always handle fishing tackle responsibly and never leave it lying on the ground as someone may trip and fall on it, step on a hook, or even break the tackle. The hooks and lures should be removed from the line and stored in the tackle box when moving the equipment. While fishing one must make sure to look behind before casting so that the hook will not catch a power line, tree, or another person.

Fig. 18.1 Angler using sun protection while fishing. (Picture credit: Abrar Khan)

1.8 Dressing Properly

When going out on a fishing trip, one should always dress for the weather and be prepared for sudden changes. Lightweight, UV blocking clothes are a must as they provide one with sun protection benefits in addition to comfort. One should also pack some extra layers in order to stay comfortable throughout the day in case it turns windy or cold. If one plans to be lounging on board, a full coverage swimwear like a rash guard, swim shirt, board shorts, and swim dress should also be carried along (Fig. 18.2).

Fig. 18.2 Properly dressed angler. (Picture Credit: Ubaid Wani)

1.9 Wearing the Right Shoes

Proper shoes appropriate to the conditions should always be worn whether fishing on shore, in a boat, or wading in the water. Stray hooks, pieces of glass, sharp rocks, and other objects on shore and in the water could cut through one's bare feet. In order to avoid that, it becomes immensely important to have the right footwear. In a boat, shoes designed to keep one's feet from slipping could help prevent a person from any mishap including drowning in water.

1.10 Carrying the Proper Accessories

Choosing the right accessories which include UV protected sunglasses, hat, personal flotation device/life jacket is critical to ensure safety to the angler. Sun exposure can cause serious sunburns on eyes, damaging the corneas as well as whites of the eyes. Protecting the eyes using the right kind of sunglasses is very important. Wraparound sunglasses with dark lenses that offer 100% UVA/UVB protection are the best options. One can also consider investing in polarized sunglasses which neutralize glare and help reduce reflection of the water when one is sight fishing. Another important accessory to be considered carrying along on a fishing trip is a wide brimmed hat that can protect the face, neck, and shoulders from sun's rays. Hats also help to keep the head cool in summer and warm in winter. When fishing from a boat, one should always carry a personal flotation device/life jacket and make sure each passenger wears one too. To be on the safer side, one should also bring extra items such as flashlights, maps, radio, GPS device, and insect repellants, etc.

2 Conclusion

Angling can be a delightful experience but at the same time it can pose some serious risks as well. Therefore personal safety must be given the top priority when one plans to go for fishing. Proper safety measures save lives, prevent serious accidents and injuries. Everyone who likes fishing should be trained in personal survival techniques, use of life jackets, fire extinguishers, emergency procedures, first aid basics, etc. Emergency smoke signals, radar reflectors can also be carried on board as additional emergency materials. With proper training and a responsible approach, angling can be the best experiences one can have in a lifetime.

Rod and Reel

19

Abstract

Choosing the appropriate rod and reel set is essential for a productive and pleasurable sport fishing trip in India. Various options are available to anglers based on the species they are targeting, the fishing conditions, and their own tastes. Fly fishing enthusiasts like to use lightweight, delicate fly rods with precision-engineered fly reels for freshwater game species like trout and mahseer. These configurations enable precise casting in confined places and delicate presentations of the fake fly. Conversely, spinning rods and reels are adaptable choices that work well with both freshwater and saltwater species, providing the strength and control required to bring larger fish under control. For targeting powerful marine game fish, anglers often rely on heavy-duty trolling and big game rods paired with high-capacity, drag-equipped reels. These robust setups can withstand the intense runs and jumps of these pelagic predators, ensuring an exhilarating yet manageable angling experience.

Keywords

Fly rods · Trout · Mahseer · Trolling · Casting · Fly

1 Introduction

A fishing rod is a specialized equipment used by fishermen to cast lines and work lures or baits in an effort to capture fish. It is an essential part of rod and reel fishing, providing the necessary control, adaptability, and sensitivity to identify bites and successfully land fish. In order to accommodate various fishing techniques and target species, fishing rods are available in a wide range of styles, materials, lengths, and actions (Fig. 19.1).

© The Author(s), under exclusive license to Springer Nature Singapore Pte Ltd. 2024

S. T. Mushtaq et al., *Angling in India*,
https://doi.org/10.1007/978-981-97-8876-7_19

Fig. 19.1 Fishing rods. Picture Credit: Ubaid Wani

2 The Elements of a Fishing Rod

2.1 Blank

The blank is the term for the fishing rod's primary body. Typically, fiberglass, graphite, or a composite material combining the two are used to make it. The rod's overall weight, strength, and sensitivity are all determined by the blank. Fiberglass rods are more resilient and have a more forgiving action, but graphite rods are lighter and more sensitive.

2.2 Handle and Grip

The back of the fishing rod is referred to as the handle, grip, or butt. For the angler, it offers a pleasant and ergonomic grip. Cork, EVA foam, or a mix of materials are frequently used to make handles. Longer handles provide you more leverage when fighting bigger fish and enable two-handed casting.

2.3 Reel Seat

The reel seat is a component located on the lower part of the rod's handle. It secures the fishing reel in place and provides stability during casting and retrieving. Reel seats are typically made of materials like graphite or aluminum and feature adjustable mechanisms to accommodate different reel sizes.

2.4 Guides

Guides, also known as line guides or rod eyes, are small metal rings that run along the length of the fishing rod. They guide the fishing line and prevent it from coming into contact with the blank, reducing friction and enabling smooth line movement during casting and retrieval. Guides can be made of materials like stainless steel, ceramic, or titanium.

2.5 Tip-Top Guide

The tip-top guide is the final guide at the top of the rod. It is slightly larger and is designed to withstand the stress and strain caused by the line during casting and fish fighting. Tip-tops are often made of durable materials such as ceramic or stainless steel.

3 Length and Action

The length and action of fishing rods vary to accommodate various fishing techniques and target species. Length usually varies from 4 to 14 feet (1.2 to 4.3 m), measured from the butt to the tip of the rod. Shorter rods offer better control in confined settings, while longer rods typically yield longer casting distances. The rod's bending and flexibility qualities are referred to as its action. There are three types of rods: medium action, slow action, and fast action. Fast action rods have a higher bending angle and can set hooks more quickly since they bend more in the top third or quarter of the blank. The middle bend of medium action rods strikes a compromise between flexibility and sensitivity. Slow action rods bend throughout the majority of the blank, offering greater forgiveness and control for delicate presentations.

4 Fishing Techniques and Rod Types

Different fishing techniques require specific rod types. Some common rod categories are:

4.1 Spinning Rods

Spinning rods are popular for their versatility and ease of use. They are equipped with spinning reels and designed for casting light lures or baits. These rods have large guides, a smaller reel seat, and a more flexible tip to accommodate the spinning reels mechanics.

4.2 Baitcasting Rods

Baitcasting rods are paired with baitcasting.

sting reels and are well-suited for precision casting and heavier lures. They have smaller guides and a stronger backbone to handle the increased weight and resistance generated by baitcasting reels.

4.3 Fly Rods

Fly rods are specifically designed for fly fishing, where the weight of the fly line is used to cast lightweight artificial flies. They are long, flexible rods made of materials like graphite or bamboo. Fly rods are categorized by line weight, and their action can range from slow to fast depending on the fishing style.

4.4 Telescopic Rods

Telescopic rods are collapsible rods that can be extended or retracted for easy transportation and storage. They are popular among traveling anglers or those who need a compact option without sacrificing performance (Fig. 19.2).

Fig. 19.2 Telescopic rod. Oleg Bor, CC BY-SA 4.0 <https://creativecommons.org/licenses/by-sa/4.0>, via Wikimedia Commons

5 Choosing the Right Fishing Rod

The right fishing rod selection is influenced by a number of variables, including the target species, fishing location, favored methods, and individual preferences. To guarantee the best possible fishing experience, the length, action, power, and material of the rod must all be taken into account.

To suit a variety of budgets, fishing rods come in a range of pricing points. Expensive rods can give better performance, sensitivity, and longevity, but less expensive models are a wonderful place to start for novices or infrequent fishers. Don't forget to take into account the local laws, fishing rules, and requirements for your particular fishing experiences. Rods can have a longer lifespan and operate at their best with proper maintenance, such as washing them in fresh water after using saltwater.

6 Fishing Reel

A fishing reel is a mechanical device attached to a fishing rod that allows anglers to store, release, and retrieve fishing line (Fig. 19.3). It is a crucial component of rod and reel angling, providing control, line management, and drag capabilities to land fish successfully. Fishing reels come in various types, sizes, and designs, each suited for specific fishing techniques and target species.

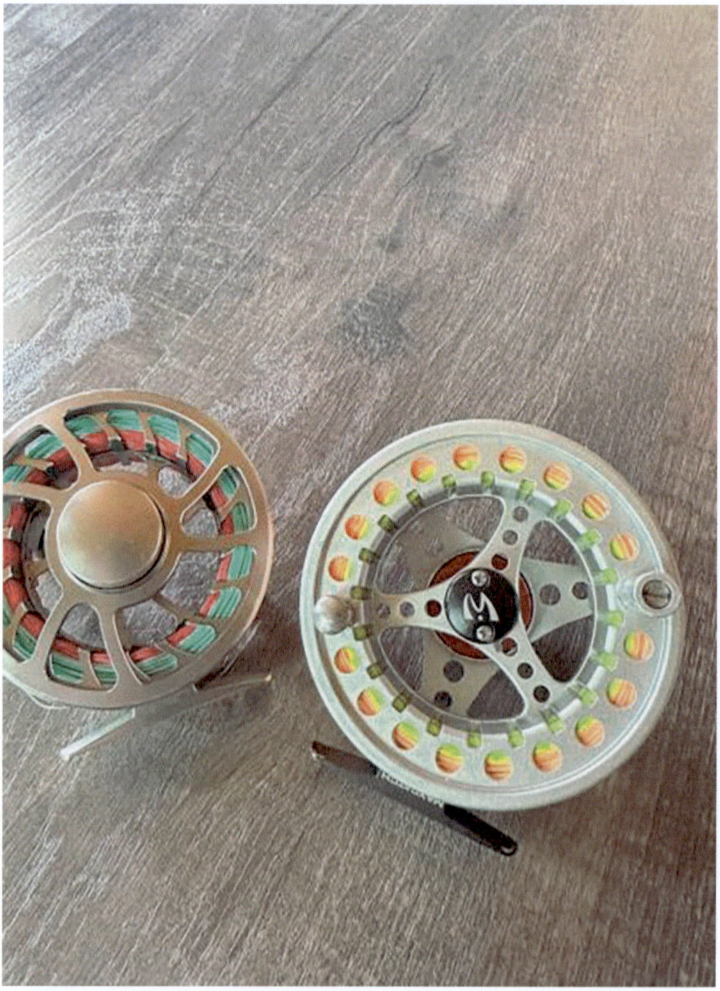

Fig. 19.3 Fishing reels. Picture credit: Ubaid Wani

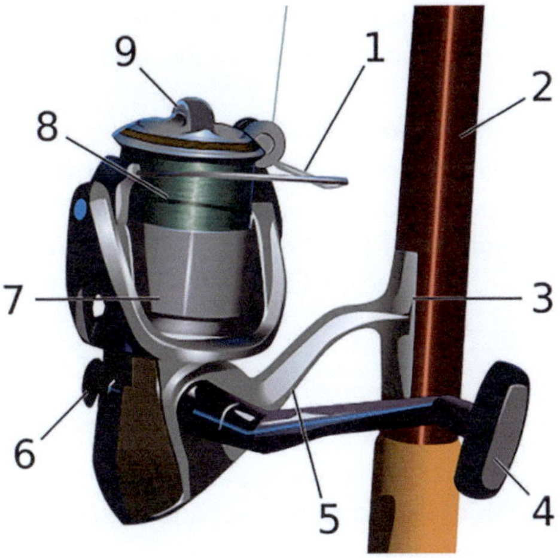

Fig. 19.4 Components of a fishing reel. (1) Pick up or bail; (2) Reel seat; (3) Reel foot; (4) Handle; (5) Support arm; (6) Anti-reverse lever; (7) Skirted spool; (8) Fishing line; (9) Drag adjustment knob. Original photo: Jan Tik. Vectorized version: Chabacano, CC BY-SA 3.0 <http://creativecommons.org/licenses/by-sa/3.0/>, via Wikimedia Commons

6.1 Components of a Fishing Reel

6.1.1 Reel Body

The reel body, also known as the reel housing or frame, is the main structural component of the fishing reel. It houses the internal gears, bearings, and other mechanisms. Reel bodies are typically constructed from materials like aluminum, graphite, or a combination of both to provide strength and reduce weight (Fig. 19.4.)

6.1.2 Spool

The spool is a cylindrical component of the reel that holds the fishing line. It rotates during line retrieval and release. Spools are usually made of materials like aluminum or graphite and come in different sizes to accommodate varying line capacities.

6.1.3 Handle

The handle is a crank located on one side of the reel. Anglers use the handle to retrieve the fishing line by turning it in a clockwise or counterclockwise direction. Handles can be single or double, and they are often equipped with ergonomic grips for comfortable operation.

6.1.4 Drag System

The drag system is a mechanism within the reel that provides resistance to the fish's pulling force. It allows the line to be released smoothly during a fish's run while maintaining tension to tire the fish and prevent line breakage. Drag systems can be front drag, rear drag, or center drag, and they typically consist of drag washers, drag knobs, and a drag adjustment mechanism.

6.1.5 Bail or Line Guide

In spinning reels, the bail is a wire arm that flips over the spool to guide the line during casting and retrieve. It helps prevent line tangling and offers line management. In bait casting reels, a line guide or level wind mechanism ensures even line distribution across the spool during retrieval.

6.1.6 Gear Ratio

The gear ratio determines the speed at which the spool rotates during line retrieval. It is expressed as a numerical ratio, such as 6.2:1, where the first number represents the number of spool rotations for each full handle turn. Higher gear ratios provide faster line retrieval, ideal for techniques that require quick lure presentation or covering large areas.

6.2 Types of Fishing Reels

Different fishing techniques and target species require specific reel types. Here are a few common reel categories:

6.2.1 Spinning Reels

Because of their simplicity of use and adaptability, spinning reels are very popular. They feature a fixed spool orientation and are positioned beneath the fishing rod. Spinning reels are appropriate for both freshwater and saltwater fishing, among other types of fishing. Their casting distance, line capacity, and smooth line release are well-known attributes. The drag adjustment is located on top of the spinning reel's open-face design, in contrast to the spincast reel. The metal bail that comes with it is meant to lock the line and keep it from unspooling. In addition, the bail plays a crucial role in uniformly guiding the line back onto the spool. The distinctive feature of spinning reels is that they fasten to the rod from below offering a comfortable gripping position in addition to a good balance during casting.

To cast with a spinning reel, the bail is disengaged, and the line is squeezed against the rod with your index finger to prevent it from unspooling. Next, swing your rod from the side or overhead. As you do, release your index finger around halfway through the motion. Aim the tip of the rod where you want the bait to land. One thing a lot of people get wrong with the spinning reel is reengaging the bail after the cast. Most spinning reels close the bail automatically after you start reeling. The first spin can often cause the line to miss the spool, resulting in a tangle. When you throw the line out, always make sure to put the bail back in its starting position by hand.

Spinning reels are an immensely capable piece of fishing equipment. They work equally well with lures and smaller baits, and are a good choice for a number of species and habitats. Combined with thin, powerful braided lines, they can produce some serious pulling power, too. If you're not careful with handling the bail, you can easily get nasty line twists and tangles. The other drawback is that you're limited to lighter gear. And once you start loading spinners with heavier lures and lines, their performance starts to drop pretty noticeably.

6.2.2 Baitcasting Reels

Baitcasting reels are commonly used for targeting larger fish species or techniques that require precise lure placement. Baitcasting reels offer excellent casting accuracy and control, particularly when using heavier lures or fishing in heavy cover. Baitcasting reels have a lot more moving parts than the spincast or the spinning reel. As a result, they come with a learning curve, but mastering them ultimately takes your fishing game to the next level.

The baitcaster is mounted on top of the rod handle. It features a semi-enclosed design, and a much sturdier build. On top of the drag mechanism, which sits next to the reel handle, the baitcaster has two additional components that allow for extra performance and customization. These are the spool tension knob, and the braking system. In essence, both mechanisms are used to adjust how fast the line is going out of the reel. It has two advantages, one gets to throw the line exactly as far as one needs to, and it also prevents the spool from turning faster than the line is going out. Baitcasting reels don't have a bail, so to stop the line from spooling, one actually has to press his/her thumb against the spool. Baitcaster rod guides are a lot smaller than the ones on a spinning setup. This is because baitcasting reels release the line in a straight path. On a spinning reel, the line goes out in a circular motion, and that requires more space.

Baitcasting reels are the most powerful fishing reels. They can handle heavy lines and produce a lot of pulling power, making them a great option when you're chasing larger fish. Baitcasters are highly customizable. However in using a baitcaster, different weights of lures demand different settings for the spool tension and braking system, which means adjusting the settings every time you change your lure. Another drawback is that baitcasters are the most expensive fishing reels.

6.2.3 Spincasting Reels

Spincasting reels, also known as closed-face reels, are beginner-friendly and easy to use. They have a closed face or cover that helps prevent line tangles. Spincasting reels are ideal for beginners, children, or anglers looking for a hassle-free fishing experience. Spincasters feature a metal nose cone, which hides all the important components of the reel. On the back, there's a button which serves to toggle the line between free-spool and locked. These reels also have a drag adjustment mechanism. This mechanism essentially allows one to adjust how much resistance a fish feels when pulling on the line. On a spincast, the "drag" is usually on the side of the reel, or next to the reel handle. Casting with a spincast reel is very easy. All one needs to do is press the spool control button, take the swing, and release. Once the button is released, the line will fly out to where the rod tip is pointing. When one wants to stop the line, the button is to be pressed one more time.

Two main advantages to using a spincast reel is that they are very simple to operate and rarely cause line tangles. Also they are the cheapest type of fishing reel. However they also have their own disadvantages such as their closed-face design tends to keep water and debris inside the reel, which can damage it over time. Also most spincasters aren't that well-made and will rarely last longer than a single season. Besides they also have limited casting range and aren't as precise as other reel types.

6.2.4 Trolling Reels

Reels made specifically for trolling are used in a fishing method in which bait or lures are pulled behind a moving boat. Generally speaking, they are bigger, able to accommodate thicker lines, and equipped with a line counter to monitor lure depth. For handling large game fish, trolling reels have strong construction and effective drag systems. These reels not only enable you to cover more ground, but they also have a lot larger spool capacity and frequently have a line counter that lets anglers determine the precise depth at which the fish are holding. Trolling reels that are robustly built and able to withstand the torque that different types of fish may put them through should be used. The least expensive and perhaps greatest choice for novices is graphite. Cast aluminum or a machined aluminum casing is ideal for anglers pursuing larger fish species like trout, salmon, or saltwater fish like tuna since it is less likely to break apart early or give up on you in the middle of a fight. A smooth drag, an audible clicker, and a level breeze are other desirable qualities.

6.2.5 Fly Fishing Reels

These are one of the simplest types of reels that you can find. That's because the purpose of these reels is to hold a lot of fly line and provide a bit of backing when that large fish tags on to your fly. Having said that, there are a variety of these reels each suited for specific fishing applications. There's one for freshwater fishing, another for saltwater fly fishing and also according to the species of fish that you are targeting. Might sound a little overwhelming at first. But you can narrow down on the right one pretty easily. There are three types that you need to be aware of. There's the standard arbor or SA reel that has the classic design and is best suited for small fishing rods. These have a very narrow reel diameter and offer more memory. But retrieval can be a chore. It is better suited for winching the fish in. Medium arbor reels are best suited for beginners as these offer a convenient middle ground. The backing is decent and retrieval is not painfully slow. Large arbor reels are the best of the lot. These offer excellent retrieval times and are easier to control. The only possible downside is the added weight and the steep price tag. But those are just tiny quibbles for someone looking for a reliable fly reel. Along with the arbor style, check the drag system. There are two types of drag systems that are used on fly reels. There's the click drag system, best suited for freshwater applications and small fish and the disc drag system better suited for saltwater and larger fish species (Fig. 19.5).

6.2.6 Centerpin Reels

This remarkable fishing reel is a simple spool on a shaft that runs on its own center pin without a drag system. It is a free spinning reel that allows the current to pull the bobber. Most beginners use the hand as the line guide to feed the line as per the demand. Instead of using a conventional drag system, you will use your thumb to control the fish. Centerpin reels are one of the most difficult reels to use. Also, you

Fig. 19.5 Fly fishing reels. Picture credit: Dr. Musadiq Allaqaband

need a much longer rod to cast these reels. 12–15 feet is the normal size. These are also the easiest reels to maintain without loops and tangles. Also, you can maintain an unrestricted movement of the bait and float without the latter even touching the water. Centerpin fishing is best suited for large game fish like king mackerel, kingfish, carp, and steelheads.

6.2.7 Surf Fishing Reels

When it comes to surf fishing, you need a reel that has a sufficiently large capacity, casts accurately with weighted baits, has a durable drag system that can tackle fighting fish, and is constructed to withstand harsh weather conditions. Sounds like too much to ask. You can use both spinning as well as baitcasting reels for surf fishing. Each one has their own set of advantages and presents a set of challenges that the angler must work around. We personally prefer the baitcaster over the spinning reel for surf fishing. It outperforms the spinning in terms of accuracy, crank power, drag handling, and overall strength.

6.2.8 Offshore Reel

These reels are designed to endure some of the toughest fishing conditions and some of the most savage fish species that the water can throw at you. Yellowfin tuna, Bluefin, amberjacks, Blue marlin, Black Marlin, Sailfish, and Albacore, you are talking about the best of the game fish world. The reel should have enough maximum drag, good quality inside parts, and tremendous casting power. Traditionally, offshore fishing has always been the baitcaster's domain as it perfectly suits most of the offshore fishing styles like casting poppers, bottom fishing for snappers, rigging live baits, jigging as well as trolling.

7 Choosing the Right Fishing Reel

Selecting the appropriate fishing reel depends on several factors, including fishing style, target species, preferred line capacity, and personal preference. Consider the reel's size, gear ratio, drag system, and overall durability to ensure it can handle the demands of your fishing pursuits. It is important to match the reel with the appropriate fishing rod, ensuring compatibility in terms of weight, balance, and fishing technique. Proper maintenance, such as regular cleaning and lubrication, helps extend the reel's lifespan and ensures optimal performance.

Fishing reels are available in a range of price points, with higher-end models offering advanced features, smoother operation, and enhanced durability. Beginners or occasional anglers can opt for more budget-friendly options that still provide reliable performance. Understanding the specific fishing regulations and requirements in your area is crucial, and obtaining the necessary licenses and permits is essential before engaging in fishing activities.

8 Maintenance and Repair of Rods and Reels

Maintenance and repair of equipment used for angling is important to protect the same and make it available for reuse. These are crucial aspects of keeping fishing rods and reels in optimal condition. Proper maintenance includes regular cleaning, inspection for damage, and appropriate storage. Timely repairs help address any issues that may arise, ensuring the continued functionality and longevity of the equipment. By taking care of fishing rods and reels, anglers can enjoy reliable performance and maximize their fishing experience.

8.1 Rods

One should be careful while transporting and storing rods to prevent damage. Many a rod has been broken in a vehicle door, trunk or tailgate, or accidentally stepped on. These should be inspected after every use and looked for frayed guide attachments, damaged guides, damage to reel seats and handles. Rods should be washed with warm mild soapy water and rinsed thoroughly. After use, the rods should be dried fully and then stored in open air or enclosed in a case. Rod tip replacement kits are available at sporting goods stores and are easily glued into place.

8.2 Reels

Reels need to be cleaned at least once a year or more, depending on how often and where they are used. To clean and lubricate, the reels should be removed from the rods. The reels should be stored only after they are clean, well lubricated, and with the drag loosened, thus preventing the washer drag systems from flattening out and failing to work properly. Never use water pressure to clean a reel as this will force contamination into critical parts of the reel.

8.2.1 Two Common Reasons a Reel Will Not Work Properly

8.2.1.1 Loops or Tangles inside the Reel

To fix this, pull line out of the reel. Often the line can be pulled out and will feel "tough" or "rough" to pull out until it gets past the tangle or loop, then it will feel easy to pull out. Next, turn the reel's handle to draw the line back into the reel—use your other hand to hold the line with some tension so that the line reels back in smoothly.

8.2.1.2 The Line Has Too Much Tension on It

Inside the reel is a tiny metal pin—a "release pin" and if the line is putting too much pressure on it, the pin can't move and release the line. To fix, simply remove the reel cover, take the line off the metal pin on the edge of the metal spool/drum and then replace the cover.

8.3 Line Replacement

All lines degrade over time and use. Exposure to ultraviolet rays of the sun will hasten the degradation of line. Replace line once a year. Choice of which line to use, composition, weight, color, etc., is dependent on the type of reel, species of fish sought, type of structure, or environment to be fished. For most freshwater fishing, use 4.5 kg, for saltwater use 9 kg line. Over-spooling of line is a major cause of reel malfunctions. Be careful not to put too much line on the reel. To replace line on a reel, remove the cover and the old line from the reel. Pass the new line from the spool so that it comes on through the line hole in the cover, and then tie it on the reel's spool. Many knots will work, but the Arbor Knot works best. Then turn the handle to wind line onto the reel, straight from the spool. The spool can be held by another person, or even left on the floor as you spool line onto the reel. When the reel is full, cut the line, and thread that cut end of the line through all the eyes on the rod and tie on tackle. Remember to recycle the old line.

8.4 Tackle

Hooks, swivels, lures, pliers, and clippers can rust if stored while damp. Leave tackle boxes open until the contents are dry. Certain plastic lures react when stored together or exposed to extreme heat, they melt and can become a glob. It is best to store different soft plastics separately, in their original packaging away from heat.

9 Conclusion

To get the most out of sport fishing in India's varied aquatic environments, choosing the correct rod and reel combination is essential. Anglers can follow their passion and help ensure the sustainable management of these priceless natural resources by acquiring the necessary equipment. The performance of an angler and the overall enjoyment of a sport fishing excursion can be greatly influenced by the selection of rod and reel. The correct equipment may improve the angling experience and aid in maintaining the biological balance of India's waters, whether you're after the powerful power of big game fishing or the delicate precision of fly fishing.

Sport Fishing Gears and Equipment

20

Abstract

Freshwater ecosystems in India are crucial habitats for a diverse array of fish species, but they face numerous challenges due to human activities and habitat degradation. These ecosystems, although occupying only a fraction of global water resources, support a significant portion of known fish species, highlighting their ecological importance. However, habitat destruction, pollution, and overexploitation have led to alarming declines in freshwater biodiversity, with more than 20% of evaluated fish species now considered threatened. Recreational fishing, deeply embedded in Indian angling culture, is encountering obstacles such as limited scientific understanding, gender disparities in participation, and inadequate regulatory frameworks. The traditional fishing techniques in the trans-Himalayan region highlight the importance of indigenous knowledge in sustainable resource management. Fly fishing, a prominent angling method globally offers diverse opportunities for trout and mahseer fishing in Indian waters. However, challenges such as declining fish populations and inadequate fishing gear documentation necessitate comprehensive management strategies. Sustainable fishing practices, including selective gear usage and habitat conservation, are crucial for preserving India's rich freshwater biodiversity and ensuring the longevity of recreational fishing activities. The evolution of fishing equipment and methods spans centuries of experimentation and refinement. From primitive tools like nets and spears to modern artificial lures and advanced tackle, anglers have continuously adapted their gear to improve their chances of catching fish. Therefore, safeguarding India's freshwater ecosystems and promoting sustainable recreational fishing practices necessitate a combination of scientific research, inclusive participation, effective regulations, and the preservation of traditional knowledge. By addressing these challenges and implementing comprehensive management approaches, India can protect its valuable

© The Author(s), under exclusive license to Springer Nature Singapore Pte Ltd. 2024
S. T. Mushtaq et al., *Angling in India*,
https://doi.org/10.1007/978-981-97-8876-7_20

freshwater biodiversity and sustain the enjoyment of recreational fishing for generations to come.

Keywords
Freshwater ecosystems · Fish species · Habitat degradation · Recreational fishing · Sustainable management

1 Introduction

India is acknowledged as one of the world's foremost mega biodiversity centers, boasting two significant biodiversity hotspots—the Western Ghats and the Eastern Himalayas—housing over 2200 finfish species (Kumar et al., 2018; Sharma et al., 2013). Among these, 73 species are categorized as cold-water finfish, constituting 3.32% of the total species (Das & Pandey, 1998). The mahseer, encompassing *Tor* and *Neolissochilus* species, is esteemed as one of the most crucial game fish globally and is designated as a high-value conservation species in freshwater systems. The climatic conditions in India are suitable for their survival. Despite their past abundance, mahseer populations have experienced a decline in both numbers and sizes in natural waters. MacDonald (1948) highlighted the sport and fighting nature of the fish in his famous work *The Rod in India and Circumventing the Mahseer and Other Sporting Fishes*. Additionally, mahseer is highly valued for its food quality, with a potential to reach sizes of up to 54 kg (Froese & Pauly, 2003). Mahseers are typically found in the "Tor zone," which refers to the altitude range of 600–1200 m in glacier-fed Himalayan Rivers (Sarma, 2009; Singh & Kumar, 2000). They also have a broader distribution in the lower reaches of rivers in the Indian peninsula (Ajithkumar et al., 1999). Mahseer is often referred to as the "King of Indian Aquatic Systems" within the Indian sub-continent (Langer et al., 2001). The majestic mahseer thrives in fast-flowing rocky streams distinguished by crystal-clear freshwater and high oxygen levels. As the largest Cyprinid, mahseer holds significant importance as both a food source and a game fish in India (Ng, 2004), representing a cherished aspect of our national heritage (Oliver et al., 2007).

Trout farming in India primarily flourishes in the upper Himalayan region and the Western Ghats due to climate constraints, with Kashmir valley playing a pivotal role (Das et al., 2019; Sarma et al., 2018). Other states such as Himachal Pradesh, Tamil Nadu (Nilgiris), Uttarakhand, Sikkim, and Arunachal Pradesh also host trout farming operations, but Jammu and Kashmir and Himachal Pradesh are leading the way (Ayyappan et al., 2006). Trout farming not only provides stable incomes but also employment opportunities, particularly in regions with limited job prospects (D'Agaro et al., 2022; Mahale, 2021; Woynarovich et al., 2011). Introduced in Europe in the early twentieth century, trout farming has become a recognized commercial and recreational venture in Kashmir. Known for their superior genetics and health benefits, Kashmiri trout, especially their eggs, are highly sought after (Ayyappan et al., 2006). With 583 private and state-aided trout farms, producing 650 tonnes of trout annually, Kashmir has emerged as a significant hub for trout farming. The province's abundant rivers, streams, and lakes provide an ideal environment for trout breeding and rearing. Cold, clear, fast-flowing streams and high-altitude lakes offer optimal conditions for trout survival (Sodhi et al., 2013). The Rainbow trout, scientifically known as *Oncorhynchus mykiss*, holds the title of the most widely distributed and farmed species globally. Originating from the Pacific and North American drainages, it now inhabits waterways across every continent except Antarctica, owing to its adaptability to various aquatic environments. It is prized both as a sport fish and a common subject for experiments (Ortega & Valladares, 2015). Trout farming is prevalent in highland catchments across tropical and subtropical regions in South America, East Africa, and Asia, with Norway, Iran, Chile, Japan, Peru, North America, and Australia leading in production (SOFIA, 2022). Trout farming not only ensures higher income but also provides employment opportunities through angling tourism, restaurants, and related services (Singh, 2015; Woynarovich et al., 2011).

The understanding of fishing gear, crafts, and methods is pivotal for the scientific and judicious management of fishery resources. In the trans-Himalayan region, traditional fishing techniques are predominantly used by inland fishermen, yet detailed documentation of indigenous methods is lacking, hindering effective management efforts. Fishing techniques are influenced by behavioral characteristics and microhabitats of fish species (Raju et al., 2016), with various types of nets and gears employed for capturing fish of different sizes (Azam et al., 2014). However, many fishing crafts in India remain primitive and non-mechanized (Khanna & Singh, 2003). Understanding traditional methods is crucial for modern improvements (Mundy & Campton, 1991), especially in the face of overfishing and declining fish populations (Amos & Bailey, 2020). Sustainable fishing practices are imperative for maintaining fish populations and aquatic ecosystem health (Amos & Bailey, 2020). Selectivity of fishing gear plays a crucial role in fisheries management, impacting catch composition and target species sizes.

2 Different Sport Fishing Techniques

2.1 Fishing Tackle and Techniques

There exist four primary categories of fishing tackle: bait-casting, spin-casting, spinning, and fly-casting (Fig. 20.1). While they vary in popularity, ease of use, fishing styles, and spool type, the distinctions are relatively minor, except for fly-casting, where the weight of the line enables precise delivery of nearly weightless flies. Among these, spin-casting stands out as the easiest, making it highly favored by children and beginners, closely followed by spinning, which enjoys similar popularity among novice anglers but is also widely embraced by seasoned fishing enthusiasts (Ayyappan et al., 2006; Beckwith & Browne, 1958).

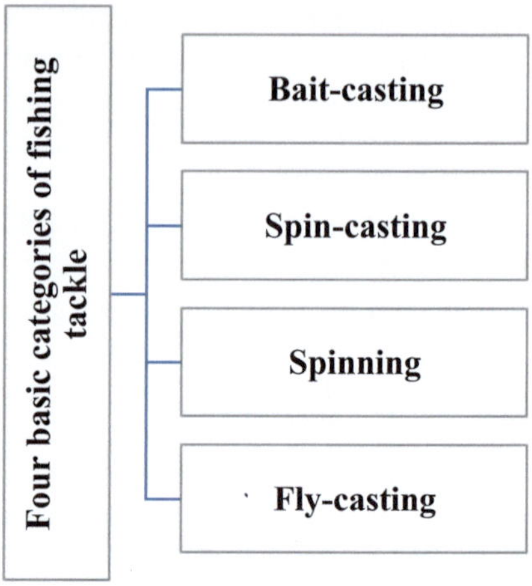

Fig. 20.1 Fish tackle categories

2.1.1 Bait-Casting

Bait-casting, a traditional fishing method, relies on live or artificial bait to lure fish and demands skill and precision, appealing primarily to seasoned anglers who appreciate its challenges. Typically, bait-casting tackle comprises a reel mounted atop a fishing rod, known for its precision and control, facilitating accurate bait placement. The reels in bait-casting setups are meticulously designed for accuracy and control, enabling anglers to cast their bait with pinpoint precision, while the rods are robust and durable, capable of handling heavy loads and providing ample leverage for casting. This gear offers versatility, allowing anglers to target a wide range of fish species in both freshwater and saltwater environments. In terms of techniques, rigging and presenting bait play a crucial role, demanding anglers to carefully prepare their bait to effectively attract fish, whether it be alive or dead, whole or cut into pieces, and presented on a hook or in conjunction with artificial lures. Additionally, mastering casting accuracy is paramount in bait-casting, necessitating anglers to control the release of a line from the reel and accurately position their bait near potential fish-holding areas, often achieved through various bait rigs consisting of hooks, leaders, swivels, and sinkers.

2.1.2 Spin-Casting

Spin-casting stands out as one of the most user-friendly fishing methods, particularly suitable for beginners and children, owing to its straightforward operation. This technique entails employing a spinning reel mounted beneath a fishing rod, featuring a convenient push-button mechanism for casting. Characterized by their simplicity, spin-casting reels require minimal skill to operate, appealing to novices seeking an easy entry into the world of fishing. Complementing these reels are lightweight and manageable spin-casting rods, facilitating the casting of lightweight lures and bait effortlessly. The versatility of spin-casting gear further enhances its appeal, enabling anglers to tackle diverse fishing environments, from tranquil freshwater lakes to dynamic coastal waters. In terms of techniques, spin-casting excels in its simplicity, with the push-button mechanism streamlining the casting process, allowing anglers to cast with ease and minimal effort. Moreover, spin-casting gear proves adept at presenting an array of lures, ranging from spinners to crankbaits and soft plastics, effectively enticing fish to strike. Given its user-friendly nature, spin-casting is often recommended as the go-to method for novice anglers eager to embark on their fishing journey.

2.1.3 Spinning

Spinning, positioned between bait-casting and spin-casting in terms of complexity, stands as a favored fishing method renowned for its adaptability and moderate skill requirement. This technique entails utilizing a spinning reel affixed below the fishing rod, available in both open-face and closed-face configurations. Distinguished by its versatility, spinning gear proves adept at addressing a broad spectrum of fishing scenarios, spanning from targeting freshwater trout to pursuing formidable saltwater gamefish. Complementing the reel, spinning rods boast longer and lighter designs compared to their bait-casting counterparts, facilitating the casting of lightweight lures and bait across extended distances. The hallmark of spinning gear lies in its user-friendly nature and capacity to accommodate light lines, rendering it an ideal choice for finesse fishing endeavors. In terms of techniques, spinning gear excels in casting prowess, capable of propelling lightweight lures and bait over considerable distances, granting anglers access to remote fish-holding areas unreachable by other tackle types. Moreover, the versatility of spinning tackle shines through in its ability to effectively present a diverse array of lures, encompassing jigs, spinners, and soft plastics, enticing fish with varied offerings. Notably, spinning gear proves indispensable for finesse fishing techniques reliant on delicate presentations and light lines, such as drop-shutting and finesse jigging.

2.1.4 Fly-Casting

Fly-fishing, a specialized technique revered for its artistry and precision, involves the use of artificial flies to imitate natural prey and allure fish. Cherished by dedicated anglers, it demands both skill and finesse to master. At its core, fly fishing relies on specialized equipment tailored to the method. Fly reels serve primarily for line storage, featuring a straightforward drag system crucial for controlling tension during fish battles. Complementing this, fly rods are characterized by their length and flexibility, enabling anglers to execute precise and delicate presentations of their flies to prospective catches (Fig. 20.2). In the realm of techniques, fly casting stands as the cornerstone of this method. It necessitates employing a specialized casting technique to propel the fly line forward with accuracy and deliver the artificial fly to the desired target area. This intricate process demands impeccable timing and coordination between the angler's movements and the rod's action. Central to the success of fly-fishing is the art of fly presentation. Anglers must adeptly mimic the behavior of natural prey, carefully selecting appropriate fly patterns and presentation techniques tailored to the target species and prevailing environmental conditions. Moreover, effective line management plays a pivotal role, as skillful manipulation of the fly line is essential for achieving casting accuracy and lifelike

Fig. 20.2 Angler with full fly fishing gear. Picture credit: Dr. Musadiq Alaqaband

presentation. From mending the line to controlling its drift and minimizing drag, mastering these aspects ensures a natural presentation that entices fish to strike.

2.2 Fishing Equipment and Fishing Accessories

The fishing equipment and methods utilized by today's anglers have evolved over centuries of experimentation and refinement. Some of the earliest fishing tackles were developed so long ago that their origins are shrouded in mystery. Primitive tools such as nets and spears were likely among the earliest fishing methods employed by early anglers, primarily focused on securing fish for sustenance. While

Fig. 20.3 Angling rods. Picture credit: Ubaid Wani

the specific origins of various fishing tackles and lures remain uncertain, their effectiveness in catching fish has endured through time. Whether it was the spoon, plug, or artificial fly that came first holds little relevance to modern anglers. The essential components of fishing tackle, from lure to rod, are arranged in a sequence that includes hook, leader, line, reel, and rod, often supplemented by accessories like sinkers and swivels (Dotson, 1980). This classification provides anglers with a wide array of options to tailor their equipment to their specific fishing needs.

2.2.1 Rods

In fishing equipment, the rod is the final component, extending from the lure. Fishing rods serve three main purposes. Firstly, they provide leverage and length for casting bait or lures, aiding techniques like fixed-line or still fishing with a cane pole. Secondly, rods extend the angler's reach, facilitating the manipulation of bait or lure in the water. Thirdly, rods guide lures into desired spots and impart motion, crucial for attracting fish. Additionally, rods absorb shock during fierce strikes and play a key role in landing fish. Rod construction varies based on the fishing type and target species, with materials ranging from split bamboo to fiberglass. Different types of fishing, such as fly-casting, spinning, bait-casting, saltwater surf, and deep-sea fishing, require specific rod materials. For instance, the universal fly rod typically measures 2.2–3.0 meters, comprising two or three pieces with spare tops. Each piece features agate or chromium steel rings at both ends and a sturdy steel ring in the center. Lightweight and flexible, these rods provide play for fish up to 500 grams and can withstand larger specimens weighing up to 10 kg (Fig. 20.3).

2.2.2 Reels

The tackle assembly sequence moves seamlessly from lure to angler, with the reel being the next component after the line (Fig. 20.4). Reels come in four main types:

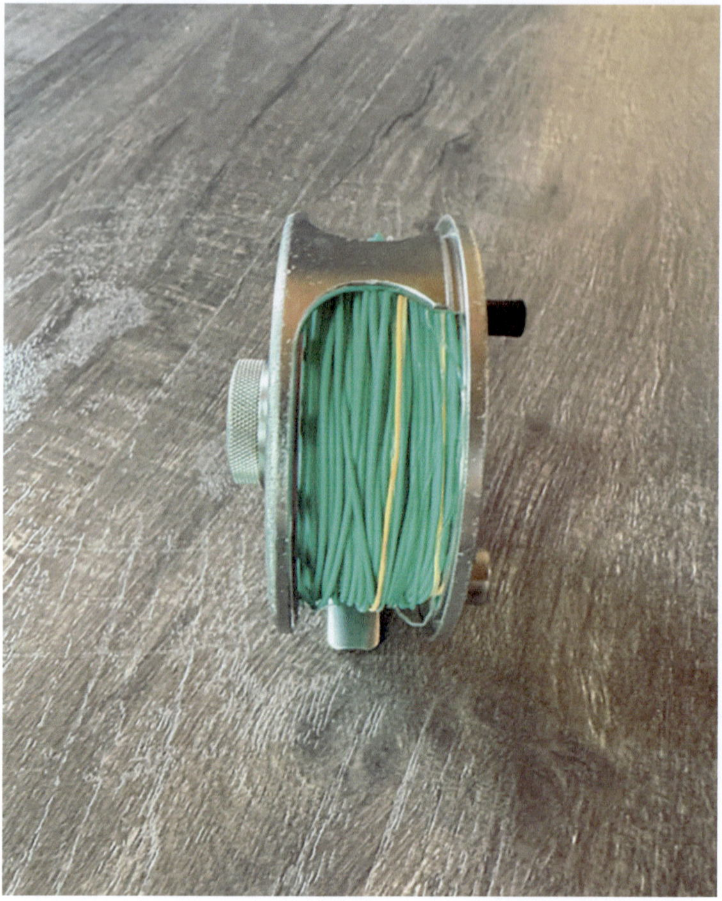

Fig. 20.4 Fishing reel. Picture Credit: Ubaid Wani

single-action, automatic, spinning, and multiplying. These types vary in construction, manipulation method, and suitability for different fishing styles.

2.2.3 Hooks

The hook, located at the end of the tackle assembly, serves as the capturing mechanism. Its optimal efficiency is achieved when the point's penetration line aligns with the traction line. The curvature of the hook, from shank to point, determines the extent of this alignment. Additionally, the shank's length and the straight portion of the hook influence its effectiveness. Hooks come in various sizes, often differing among manufacturers and even within the same manufacturer. Recent years have seen complexity in numbering and lettering systems for hooks. To simplify this, a Limerick scale provides an easier means to express the size and variety of hooks commonly used in game fishing.

2.2.4 Leaders

The leader, positioned between the fishing line and lure, reinforces the point prone to breakage during casting. Crafted from durable materials like piano wire or synthetic gut, it prevents line severance. Leaders in fly fishing come in two types: level and tapered. Tapered leaders, heavier at the line end, gradually decrease in diameter towards the lure end. They range from 1.0 to 2.0 m in length and are versatile for various fishing conditions. Level leaders maintain a uniform diameter and are ideal for “wet” fly-fishing in clear waters. They typically range from 1.0 to 2.0 m and feature dropper loops for attaching additional flies.

2.2.5 Lines

Lines for fishing are tailored to specific fishing methods, such as bait casting, salt-water fishing, fly fishing, and spinning. Salt-water fishing lines are typically made from metal or fiber, while bait-casting lines have transitioned from braided silk to nylon. Despite their pound-test rating, fishing lines can handle catches exceeding their stated weight due to various factors like rod flexibility, angler handling, water friction, and line stretching. Braided silk, dressed enamelled, oiled silk, and nylon are among the commonly used line materials, each with its advantages and limitations. Enamelled lines, for instance, are unsuitable for tropical climates due to their tendency to become tacky and brittle. Lines are graded from “A” to “H” based on their strength, with “A” being the strongest. In fly-fishing, “D” grade lines are recommended for their ability to carry lightweight lures effectively. Tapered lines, graded from “C” to “E,” transition from heavier to lighter weights. Additionally, tubular lines that float have gained popularity for their improved visibility.

2.2.6 Tackle Accessories

In addition to the essential fishing tackle like lure, hook, leader, line, reel, and rod, anglers often rely on certain accessories to enhance their fishing experience. These accessories include swivels, which prevent line twisting; sinkers used to help lures sink deeper; and snaps, functioning as safety pins to quickly attach and release lures. Another unique fishing rod described by Goldschmidt (1957) features a 75 cm length made from ordinary bamboo with a naturally grown pistol grip butt. Attached to this is a 40 cm tip made from a buffalo horn, tapered down to a fine point. The rod is used in Hyderabad and is capable of landing fish weighing up to 50 lb. (22.68 kg) from local tanks. The line is wound on a wooden reel similar to those used for flying kites, passing through a small hole in the rod’s butt and a silk loop at the tip.

2.2.6.1 Handling the Tackle

Proficiency in handling sport fishing tackle requires practice and skill. The goal is to effectively manipulate the tackle to attract fish. The fundamentals of bait and fly casting manipulation are closely intertwined, with techniques like wrist movement, forearm motion, rod spring on the back cast, and motion points being similar in both types of fishing.

2.3 Artificial Bait

2.3.1 Lures

The primary component of fishing tackle, the lure, is divided into three main subdivisions: (a) spinners and spoons, (b) plugs or lures with bodies, and (c) flies (Mahanta et al., 2009; Modassir, 2011; Sehgal, 1987). Each of these subdivisions is further divided into various types of devices, all designed to attract and catch fish. Below, the construction and types of each of the three subdivisions are described in detail.

2.3.2 Spoons and Spinners

Spoons and spinners come in various types, including spoons that wobble on retrieval and spinners that whirl. This combination of lures is among the oldest artificial lures, known for its simplicity in basic construction. The principle behind its design is that the blade, typically made of metal or other materials, creates resistance against the flowing water as the lure moves. The simplest type is the wobbler, which serves as a universal lure favored by anglers of all levels. A wobbler spoon's fundamental component is a dished blade, which can be either symmetrical or asymmetrical along its central axis. Typically, there is a hole in the forward end for fitting a split or swivel, with the hook or hooks either solidly attached to the blade itself or free-swinging on a split ring through a hole in the rear end of the blade. The dish of the blade features a planing surface that generates action as the spoon moves through the water. Water pressure on the dished surface causes the blade to sway from side to side. In a symmetrical oval spoon with even dish depth from perimeter to center, water pressure remains somewhat consistent across the middle. A deeper dish results in wider dives and more pronounced wriggling compared to a shallower dish. The blade's attachment to the leader at one end causes it to whip and wobble from side to side during retrieval. The hook attachment distinguishes two variations of wobbler spoons: fixed hook and free-swinging hook. In a fixed hook wobbler spoon, the hook may be soldered or modified for attachment via a screw, allowing for easy removal if desired. Examples of typical fixed hook wobbler spoons include the standard wobbler and silver minnow. Conversely, a free-swinging hook wobbler features a hook attached to the lower end of the blade using a split ring, as seen in a fine lander spoon.

The principle behind the construction of a spinner involves using thin blades made of metal or other materials that rotate in a circular motion around the axis of the line of traction. Spinners typically have either a single blade attached at one end, with each blade turning independently, or multiple blades without attachment to each other. They are classified based on the attachment method and the shape of the blade. In some cases, the blade is attached directly, while the shaft is absent. In other types of spinners, the blade or blades rotate freely around a bearing on a shaft. This bearing can take various forms, such as a hole at the attached end of the spoon, an eyelet of wire soldered or welded to the end of the spoon encircling the shaft, a clavis-shaped "collar" or "saddle," or a hole or loop provided in the upper end of the spoon blade. Spinners can also come in multiple series, where several spoons, simple blades, or shafts are linked in a line based on the shape and type of blades used. They are further classified into various shapes such as circular, broad oval, narrow oval, shoe-shaped, egg-shaped willow-leaf, kidney-shaped, flinted, and others.

The propeller-blade spinner features double blades arranged in a propeller shape, rotating around the shaft on a bearing located in the center. It can be a simple propeller type where one or more spinners, often two, rotate on a shaft in opposite directions to counteract the twisting of the line. This type of spinner is a self-contained moving unit and is not attached to any other part of the lure. Typically, it is attached before plugs or flies in the fishing setup.

2.3.3 Plugs (Lure with Body)

The lure with a body, known as a plug, can either float on the surface or submerge underwater. Its buoyancy allows it to float when pulled by the line, and it sinks when traction is released or weights are added, enabling it to reach deeper zones of a body of water. The plug typically resembles a fish or other live bait, with the first marketed plugs initially made from wood or wire. Wooden plugs have been a traditional choice, but in recent years, plastic has emerged as a popular alternative due to its versatility. Plugs are generally classified into three categories based on their behavior in water, playing surfaces, and center of weight: (a) floating with surface disturbance, (b) sub-surface, and (c) deep-sinking.

The floating-cum-surface disturbance group comprises plugs that remain buoyant on the water's surface. They can be made to sink by adding weight or a sinker. Their primary function is to imitate struggling or moving prey on the surface, enticing strikes from fish. In terms of construction, their planing surfaces are designed to make them dart from side to side. Some floating plugs may also feature spinner or flapping blades to further agitate the water and attract attention. This group can be categorized into four types: (1) fish or small animal-like plugs, (2) plugs with spoon-spinner attachments, (3) plugs with planing surfaces causing darting or wobbling movements, and (4) popper lures resembling small fish or other body-shaped lures.

Sub-surface plugs encompass various categories based on their sinking property when traction is applied. These categories include (a) planing surfaces shaped like the lure's head, (b) planing surfaces curved in both the head and body, (c) planing surfaces featuring metal diving plate parts in the plug head, and (d) metallic plates attached outside near the throat of the plug. Deep-cum-sinking plugs are designed to wobble and follow the lines of traction without moving sideways. Propeller spinners provide the flashing movement, while planing surfaces, either head-shaped or both head and body-shaped, contribute to darting and wobbling actions. Plugs, serving as lures, find application in almost all types of sport fishing, enticing fish with their simulation of various prey items such as meat chunks, small fish, animals, or birds. They are primarily used in casting but can also be employed for trolling.

2.3.4 Flies

Flies utilized in fishing fall into two categories: "dry" and "wet." Dry flies float on the water's surface and are sometimes varnished for water resistance. Wet flies, which include nymphs and streamers, are often treated with soap or dipped in detergent to facilitate sinking, earning them the classification of "wet." Additionally, metal spoons or spinners may accompany wet flies. Fly fishing, predominantly for trout in streams, sees the use of patented spinners such as Halcyon, Phantoms, Pennell, Devons, and Fly Minnows. Among flies simulating insects or bugs, variations are abundant due to differences in their structural features, taking forms like nymphs, hackles, palmers, and spiders. Wings are categorized based on their shapes, such as closed-wing, fan-wing, split-wing, divisible, and double-wing (Fig. 20.5). Streamer flies and bucktails serve as lures mimicking minnows or other fish prey, typically dressing long hooks with long wings. These wet-fly type lures may not always attract game fish, as they closely resemble fish rather than insects. Notably,

Fig. 20.5 Various flies used as trout bait in fly fishing. Picture credit: Ubaid Wani

streamers have feathered wings, while bucktails feature wings of hair or fur from animals other than deer, sometimes topped with small feathers of peacock or ostrich. Streamers and bucktails hold appeal, particularly to larger fish, as they mimic the diet of carnivorous sporting fish, making them vital in saltwater fly fishing and useful in inland waters during spring when insect hatches are scarce. Their unique angling advantages make them indispensable in the equipment of versatile anglers, each of whom has their preferred fly patterns. Commonly favored patterns include Grey hackle in peacock, Royal Coachman, Queen of the Waters, White miller, March brown, Wickham's fancy, Blue quill, Pink lady, and Red ant. Fly fishing enjoys popularity for both trout and mahseers.

2.3.5 Miscellaneous Artificial Lures

Apart from spoons, spinners, plugs, and flies, there exist other artificial lures designed to mimic bugs, mice, frogs, fish, worms, and more. The bug lure falls between the plug and the fly, featuring a body crafted from materials like hair, wool, or silk. Additionally, genuine lures are replicating natural entities such as mice, frogs, fish, and worms using rubber and other synthetic materials, all aimed at attracting game fish (Sehgal, 1987).

2.4 Natural Baits

2.4.1 Live Baits

Live bait, a type of fishing bait, refers to living organisms used to attract and catch fish, typically attached to a fishing hook or placed inside a fish trap. Common natural baits include worms, minnows, leeches, insects, and shrimp. Unlike artificial lures that mimic fish prey, live bait consists of actual fish prey. Some fishermen also use plastic bait or electronic lures to attract fish. Studies indicate that natural baits like croaker and shrimp are often more readily recognized and accepted by fish. The choice of baiting technique depends mainly on the target species and its habitat. Natural bait anglers typically use common prey species of fish as attractants, such as worms, leeches, minnows, frogs, salamanders, and insects. Natural baits are effective due to their lifelike texture, odor, and color, which make them appealing to fish (Sehgal, 1987).

2.4.2 Worms

Worms are a widely used bait in freshwater angling, with common earthworms being a universal choice. In trout fishing, grubs, maggots, grasshoppers, bees, and ants are also utilized, although many anglers favor roe as the superior bait. In warmer climates, lake fish like bream are attracted to bite-sized bread bait. Most common earthworm species, such as *Lumbricus terrestris*, can be easily sourced from the garden and are well-suited for freshwater fishing. However, they are not ideal for worm farming due to their deep burrowing habits and reluctance to breed in shallow bins. Red compost worms, such as the red wiggler or European nightcrawler, are preferred for farming as they are surface-dwelling detritivores. The

European nightcrawler, especially sought after, is prized for its tolerance to near-freezing temperatures and suitability for saltwater fishing.

3 Conclusion

Specialized gear and equipment are needed for sport fishing in India in order to guarantee a sustainable, fun, and safe experience. The choice of equipment can have a big impact on an angler's success and the overall enjoyment of a sport fishing trip, from the lightweight fly rods and expertly designed reels used in fly fishing to the sturdy spinning rods and baitcasting setups used to target large game fish. Aside from rods and reels, other necessary gear consists of premium lines, leads, tippets, and a variety of artificial lures, flies, and baits that are matched to the target species and fishing circumstances. The use of accessories like landing nets, hook removers, and fish-friendly handling equipment is essential for encouraging catch-and-release methods and reducing stress on the captured fish. In addition to improving the fishing experience, the availability and appropriate usage of this specialist equipment helps ensure the long-term viability of India's game fish stocks. Sport fishers may make sure they can follow their passion and responsibly contribute to the preservation of these priceless natural treasures by arming themselves with the appropriate tools and tactics.

References

Ajithkumar, C. R., Remadevi, K., Thomas Raju, K., & Biju, C. R. (1999). Fish fauna, abundance and distribution in Chalakudy river system, Kerala. *Journal Bombay Natural History Society, 96*(2), 244–251.

Amos, H., & Bailey, M. (2020). Linking small-scale fishing and community capitals: The case of Atlantic cod. In *Routledge handbook of sustainable and regenerative food systems* (pp. 154–170). Routledge.

Ayyappan, S., Jena, J. K., Gopalakrishnan, A., & Pandey, A. K. (2006). *Handbook of fisheries and aquaculture*. ICAR.

Azam, S., Saha, A. K. M., Asadujjaman, K. R., & Minar, M. H. (2014). Fishing gears and crafts commonly used at Hatiya Island: A coastal region of Bangladesh. *Asian Journal of Agricultural Research, 8*(1), 51–58.

Beckwith, M. C., & Browne, E. (1958). Casting and angling a new activity in our required program. *Journal of Health, Physical Education, Recreation, 29*(5), 12–51.

D'Agaro, E., Gibertoni, P., & Esposito, S. (2022). Recent trends and economic aspects in the rainbow trout (*Oncorhynchus mykiss*) sector. *Applied Sciences, 12*(17), 8773.

Das, P., & Pandey, A. K. (1998). Current status of fish germplasm resources of India and strategies of conservation of endangered species. In A. G. Ponniah, P. Das, & S. R. Verma (Eds.), *Fish genetics and biodiversity conservation* (pp. 252–212). Nature Conservators.

Das, D. N., Abujam, S., & Singh, A. D. (2019). *Research trends on fish & fisheries in mountain waters of eastern Himalayan region*. Notion Press.

Dotson, R.C. (1980). Fishing methods and equipment of the US west coast albacore fleet.

Froese, R., & Pauly D. (2003). *World Wide Web electronic publication*. http://www.fishbase.org

Goldschmidt, P. (1957). An indigenous fishing tackle and rod. *Journal of the Bombay Natural History Society, 55*, 953.

Khanna, S. S., & Singh, H. R. (2003). *A text book of fish biology and fisheries* (p. 524). Narendra Publishing House.

Kumar, N. K., Hanchinal, R. R., Gupta, A., & Dandin, S. B. (2018). Role of biodiversity international in agricultural and horticultural research in India. *Indian Farming, 67*(10), 21.

Langer, R. K., Ogale, S. N., & Ayyappan, S. (2001). *Mahseer in Indian subcontinent-a bibliography*. Central Institute of Fisheries Education.

MacDonald, A. (1948). *Circumventing the Mahseer and other sporting fish of India and Burma* (pp. 306–307). Bombay Natural History Society.

Mahale, S. (2021). *Trout swims to success in Jammu and Kashmir*. Mongabay. http://www.mongabay.com

Mahanta, P. C., Sarma, D., & Joshi, K. D. (2009). *Sport fishing and livelihood in upland Himalayan region* (pp. 29–31). 20th All India Congress of Zoology.

Modassir, Y. (2011). Fish species diversity and its potential for angling fishing in Goa–an overview. *Indian Journal Applied & Pure Biolgy, 26*(2), 291–304.

Mundy, P., & Campton, J. L. (1991). Indigenous communication and indigenous knowledge. *Development Communication Report, 74*(3), 1–4.

Ng, C. K. (2004). *King of the rivers: Mahseer in Malaysia and the region* (p. 170). Inter Sea Fishery (M) SDN BHD, Kuala Lumpur.

Oliver, K., Sangma, N., & Basavaraja, N. (2007). Deccan Mahseer (*tor khudree*) of Karnataka-n location of its wild brooders and fry and breakthrough in the hatchery production of its seed. *Fishing Chimes, 26*(10 Everard), 32–36.

Ortega, C., & Valladares, B. (2015). Analysis on the development and current situation of rainbow trout (*Oncorhynchus mykiss*) farming in Mexico. *Reviews in Aquaculture, 9*, 1–9.

Raju, C. S., Rao, J. C. S., Rao, K. G., & Simhachalam, G. (2016). Fishing methods, use of indigenous knowledge and traditional practices in fisheries management of Lake Kolleru. *Journal of Entomology and Zoology Studies, 4*(5), 37–44.

Sarma, D. (2009). Chocolate mahseer (*Neolissochilus hexagonolepis*): A candidate fish for hill aquaculture. *Fishing Chimes, 29*(7), 8.

Sarma, D., Akhtar, M. S., & Singh, A. K. (2018). Coldwater fisheries research and development in India. In *Aquaculture In India* (pp. 93–133). Narendra Publishing House.

Sehgal, K. L. (1987). *Sport fisheries of India* (p. 125). Sport Fisheries of India.

Sharma, S. K., Rana, J. C., & Chopra, V. L. (2013). Biodiversity (plants/animals/microbes/birds): Status, endemism, threatened species. In *Climate change and its ecological implications for the Western Himalaya* (pp. 180–216). Scientific Publishers.

Singh, A. K. (2015). Advances in Indian Coldwater fisheries and aquaculture. *Journal of Fisheries Sciences, 9*(3), 48–54.

Singh, H. R., & Kumar, N. (2000). Some aspects of ecology of hill streams; stream morphology, zonation, characteristics, and adaptive features of ichthyofauna in Garhwal Himalaya. In J. S. Datta Munshi (Ed.), *Modern trends in fish biology research* (pp. 1–18). Narendra Publishing House.

Sodhi, A. S., Saroch, J. D., & Verma, J. (2013). Fisheries resources of Kashmir: A case study of river Jhelum. *Journal of Chemical Biological and Physical Sciences (JCBPS), 3*(2), 1194.

SOFIA, (2022). State of World Fisheries and Aquaculture 2022 report. 266. https://www.fao.org/3/cc0461en/cc0461en.pdf

Woynarovich, A., Hoitsy, G., & Moth-Poulsen, T. (2011). Small-scale rainbow trout farming. *FAO Fisheries and Aquaculture Technical Paper, 561*, 1.

Ecotourism and Recreational Fisheries 21

Abstract

In India, ecotourism and recreational fishing provide a special fusion of leisure pursuits and environmental preservation. A vast variety of game fish species are supported by the nation's varied environments, which span from the tropical coasts to the Himalayan rivers, drawing anglers from all over the world. Important locations include the Himalayan states of Himachal Pradesh, Uttarakhand, and Jammu and Kashmir, where trout and mahseer fishing can be found in rivers like the Pabbar Valley and the Ramganga. Marine game fish species such as tuna and sailfish can also be found in the coastal regions of the Andaman & Nicobar Islands and the Western Ghats. The construction of protected areas and catch-and-release fishing are two examples of effective conservation and management techniques that are essential to preserving the natural balance and guaranteeing the long-term viability of these resources. India can support local businesses, protect natural ecosystems, and provide tourists an exciting angling experience by encouraging ecotourism and recreational fishing.

Keywords

Ecotourism · Pabbar valley · Ramganaga · Conservation · Sailfish

1 Introduction

Recreational fishing is an integral part of the tourism industry in many countries, offering significant economic, social, and environmental benefits. Fisheries-Based Ecotourism (FbE) has emerged as an innovative approach that combines traditional fishing activities with ecotourism principles. This chapter explores the concept of FbE, its potential impacts, and the challenges associated with its implementation. It also provides a strategic framework for developing sustainable FbE practices.

© The Author(s), under exclusive license to Springer Nature Singapore Pte Ltd. 2024

S. T. Mushtaq et al., *Angling in India*,
https://doi.org/10.1007/978-981-97-8876-7_21

2 The Concept of Ecotourism

Ecotourism, as defined by The International Ecotourism Society (TIES), is "responsible travel to natural areas that conserves the environment, sustains the well-being of the local people, and involves interpretation and education. Ecotourism focuses on enhancing natural systems through tourism and supports local livelihood diversification and nature conservation sustainably (Hossen et al., 2015). It contributes to sustainable productivity, infrastructure development in remote areas, and boosting employment rates in local communities (Clifton & Benson, 2006).

3 Linking Fisheries to Tourism

Fishing is not only a vital protein source and ecosystem service provider but also offers recreational benefits (Dugan et al., 2006). As tourists increasingly seek more engaging experiences beyond traditional sun, sand, and sea packages, fishing activities and cultural exploration related to ancient fishing heritage have become significant attractions (Fisheries and Tourism Creating Benefits for the Community, 2017). FbE includes a variety of activities such as recreational spin-cast fishing, excursions on professional fishing vessels, learning alternative fishing methods and gear from locals, game fishing, and exploring exotic marine environments and their biodiversity (Marine Fisheries Related Tourism in the EU, 2017).

4 The Importance of Recreational Fishing

Recreational fishing is one of the world's most popular leisure activities and an important sector within the global tourism industry (Borch et al., 2011). Recreational fishing, or fishing for fun rather than food, is estimated to have 220 million to 700 million anglers worldwide, and it is increasing quickly. In the United States, recreational fishing generates approximately $115 billion annually and supports 800,000 jobs (Ditton et al., 2002). North American recreational fishers contribute directly to fisheries management, conservation efforts, and outdoor recreation opportunities through fees, excise taxes, and licenses (Arlinghaus & Cooke, 2009). Similarly, recreational fishing is popular in Australia, New Zealand, Asia, and Europe, contributing significantly to their respective tourism industries (Gupta et al., 2015). Recreational fishing has many potential economic benefits, but if it is not managed effectively, it can also pose hazards to conservation, including overfishing, the selective harvesting of enormous "trophy" fish, habitat destruction, and the introduction of invasive species.

5 Fisheries-Based Ecotourism (FbE)

Three main aspects make up the concept of ecotourism: an emphasis on the natural world, sustainable management, and environmental education. These features place ecotourism wholly as a part of sustainable tourism and as a subset of nature tourism. The argument against using fishing as a type of ecotourism has historically concentrated on the nature of the consequences it has on the environment rather than addressing the issue of whether or not those effects are long-term sustainable. Although the legality of catch-and-release fishing is debatable, fishing is generally considered a consumptive activity that yields products. The implication is that non-consumptive activities, on the other hand, that offer experiences rather than goods, have less of an impact and should thus be specifically classified as ecotourism.

FbE tourism is a novel proposal for both established coastal tourism destinations and hinterland areas. It aims to provide environmentally responsible and culturally sensitive travel to fishery areas and on recreational fishing expeditions within these areas (GLOBE FISH, 2018). By offering fishers an opportunity to supplement their incomes and encouraging them to maintain their traditional environments, FbE promotes sustainable growth in the artisanal fishing sector and helps preserve this way of life (Meneghello & Mingotto, 2016). Around the world, ecotourism and recreational fishing are becoming crucial parts of sustainable tourism and fisheries management. Through the non-consumptive use of natural resources, ecotourism—which is defined as responsible travel to natural places that preserve the environment and enhance the well-being of local people—can offer socioeconomic benefits.

The main goal of nature tourism is to appreciate animals or largely unspoiled natural places. There are other aspects to ecotourism. According to Goodwin, ecotourism is low-impact nature tourism that either directly supports the preservation of species or habitats through conservation efforts or indirectly does so by bringing in enough money to support the locals' need to value and safeguard their wildlife heritage as a source of income. According to Honey (1999), ecotourism is small-scale, low-impact visitation to pristine, fragile, and typically protected regions. It should also aid in tourist education, provide funding for environmental preservation, directly support local communities' political and economic growth, and promote tolerance for other cultures and their human rights. Because fishing is consumptive, most tourism academics would not classify it as "ecotourism".

There are instances where catch-and-release fishing is conducted, which offers direct financial support (via sales of local crafts, park entrance fees, and licenses) as well as beneficial economic benefits (jobs, additional revenue, etc.) for the local community. More significantly, it works to develop moral conduct that preserves and grows fish populations, offers political backing for clean water and regional conservation projects, encourages a stewardship ethic, and fosters reverence for the environment and natural areas. If fishing is viewed from an ecotourism perspective, then increasing the number of visitors at a fishing destination can optimize these advantages.

6 Economic, Social, and Environmental Impacts of FbE

Tourism, particularly in rural and developing areas, can bring numerous economic, social, and environmental benefits (Brohman, 1996). FbE tourism enhances local economies through tourist expenditures, supports local businesses, and creates employment opportunities, thereby reducing out-migration and poverty (Fig. 21.1). Social interactions between tourists and local communities foster mutual appreciation, cultural awareness, and friendships, contributing to social infrastructure improvements such as schools, libraries, and health centers (Zambrano et al., 2010).

FbE also promotes environmental conservation by recognizing the role of professional fishers in shaping coastal areas and valuing local fishery products and traditional artisanal fisheries. This recognition can lead to a reduction in over-fishing and improved conservation of fish and marine resources (Kavallinis & Pizam, 1994).

Tourist fishing licenses can also significantly increase the amount of money available to state fishery agencies, as many of them are more expensive than resident fishing licenses. These expenses inject fresh funds into the local and regional economies, broadening the economic base of the community and providing residents with an option to more consuming types of fishing. In conclusion, there are

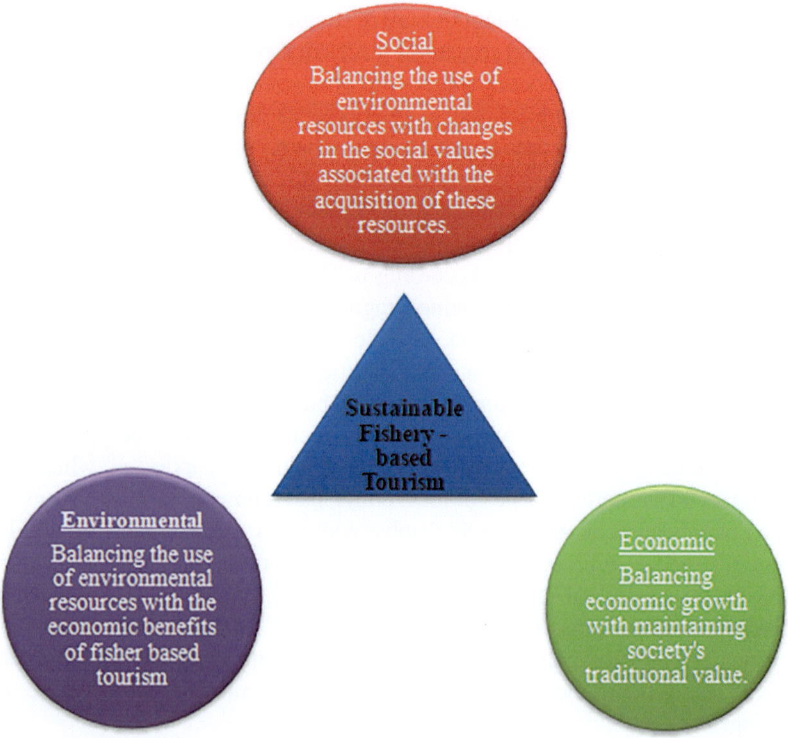

Fig. 21.1 Balancing the Pillars of Sustainable Fishery-Based Tourism

advantages and disadvantages to changing the number of tourist angler fishing days in a state or area. The quantities and proportions of nonresident fishing days that are exported between states vary greatly. Distance, cost, season, knowledge, ease and/or expense of getting a nonresident license, perceived attractiveness of fishing prospects, degree of fishing specialty, and other factors are likely to play a role in this. While there are chances for increased angling visitors that can be sustainable through the use of ecotourism, some of these elements are exogenous to management. To be achieved, though, fishery managers will have to learn more about fishing tourism and tourism in general, as well as form cooperative relationships with regional and state tourism promotion agencies.

7 Opportunities and Challenges for Ecotourism

As recreational fishing and tourism expand globally, the need to define angling ecotourism becomes increasingly important to conserve ecosystems and manage anglers' impacts. High standards benefit conservation and local economies, allowing environmentally-conscious operators to market premium services based on a fishery's environmental attributes. However, the term "angling ecotourism" risks misuse, leading to "green washing" and diminishing the value of genuine ecotourism efforts (Honey, 1999).

Increased demand can attract mass tourism, stressing and degrading regions' environmental and social infrastructure, as seen in parts of Alaska and British Columbia. Common issues include erosion, noise pollution, and improper waste disposal (Togiak National Wildlife Refuge Core Planning Team, 2001). Regions aiming for angling ecotourism need clear attractions, such as Kamchatka's steelhead or Alaska's king salmon, but must recognize logistical and cost limitations. Local markets may be more viable than competing internationally (Butler, 1990).

Successful angling ecotourism requires a comprehensive strategy, integrating diverse forms like wildlife watching, adventure trips, and ethnotourism. Early cooperation among conservation groups, regulatory agencies, communities, and operators is crucial. Thoughtfully formulated standards and guidelines are needed to transition angling ecotourism from a loosely-applied concept to a sustainable practice.

8 Management Strategy for Sustainable FbE

To develop sustainable FbE practices, a carefully planned management strategy is essential. This strategy should focus on three factors: potential sectors, information gathering through studies, and FbE governance (Sarker et al., 2018). Effective collaboration among public and private sector stakeholders, including research institutes, government agencies, donor agencies, NGOs, recreational fishing tour operators, and local communities, is crucial for the successful implementation of FbE (Bornhorst et al., 2010; Ford et al., 2012).

9 Key Elements of a Sustainable FbE Management Framework

9.1 Natural Resource Management

Ensure proper environmental management processes to preserve biodiversity and make the best use of natural resources.

9.2 Cultural Preservation

Respect the socio-cultural identity of local communities and work towards preserving cultural and traditional values while fostering intercultural understanding and tolerance.

9.3 Economic Development

Guarantee lasting economic progress by enabling stable employment and increased income opportunities for all actors involved in the tourism sector.

9.4 Legislation and Regulation

Develop and enforce appropriate legislation, standards, guidelines, and safety codes to manage FbE activities effectively.

9.5 Stakeholder Coordination

Integrate the activities of different agencies, including government, non-governmental organizations, fishers, scientists, and tour operators, to achieve sustainable tourism development.

10 Conclusion

Fisheries-Based Ecotourism (FbE) represents a promising avenue for sustainable tourism, offering economic, social, and environmental benefits while preserving traditional fishing practices and natural resources. The idea of ecotourism offers a helpful way to comprehend, assess, and lessen the different effects of nonresident anglers, even though it is frequently used to describe nonconsumptive pursuits like birdwatching and whale watching. A subset of nature tourism, which is a larger type of tourism, includes ecotourism. According to Laarman and Durst, the primary draw of nature tourism is a location that is centered around natural resources and offers activities like hiking, biking, scuba diving, and taking in the natural landscape, plants, animals, or birds. By adopting a strategic management framework and fostering collaboration among stakeholders, FbE can contribute to the sustainable development of both coastal and hinterland tourism destinations. Promotion of tourism for fishing is a successful venture and in the future more anglers are expected to travel for recreational fishing. However, it will be increasingly important to consider the social, economic, and environmental impacts of this activity.

References

Arlinghaus, R., & Cooke, S. J. (2009). Recreational fisheries: Socioeconomic importance, conservation issues and management challenges. In *Recreational hunting, conservation and rural livelihoods: Science and practice* (pp. 39–58). Blackwell Publishing Ltd.

Borch, T., Moilanen, M., & Olsen, F. (2011). Marine fishing tourism in Norway: Structure and economic effects. Økon. *Fiskeriforskning, 21*, 1–17.

Bornhorst, T., Ritchie, J. B., & Sheehan, L. (2010). Determinants of tourism success for DMOs & destinations: An empirical examination of stakeholders' perspectives. *Tourism Management, 31*, 572–589.

Brohman, J. (1996). New directions in tourism for third world development. *Annals of Tourism Research, 23*, 48–70.

Butler, R. W. (1990). Alternative tourism: Pious hope or Trojan horse? *Journal of Travel Research, 28*(3), 40–45.

Clifton, J., & Benson, A. (2006). Planning for sustainable ecotourism: The case for research ecotourism in developing country destinations. *Journal of Sustainable Tourism, 14*, 238–254.

Ditton, R. B., Holland, S. M., & Anderson, D. K. (2002). Recreational fishing as tourism. *Fisheries, 27*, 17–24.

Dugan, P., Dey, M. M., & Sugunan, V. V. (2006). Fisheries and water productivity in tropical river basins: Enhancing food security and livelihoods by managing water for fish. *Agricultural Water Management, 80*, 262–275.

Fisheries and Tourism Creating Benefits for the Community. (2017). www.webgate.ec.europa.eu/fpfis/cms/farnet/files/documents/FARNET_Fisheries_and_Tourism-9_EN.pdf

Ford, R. C., Wang, Y., & Vestal, A. (2012). Power asymmetries in tourism distribution networks. *Annals of Tourism Research, 39*, 755–779.

GLOBEFISH—Analysis and Information on World Fish Trade. (2018). Retrived May 24, 2018 from www.fao.org/in-action/globefish/publications/details-publication/en/c/1013325

Gupta, N., Raghavan, R., Sivakumar, K., Mathur, V., & Pinder, A. C. (2015). Assessing recreational fisheries in an emerging economy: Knowledge, perceptions and attitudes of catch-and-release anglers in India. *Fisheries Research, 165*, 79–84.

Honey, M. (1999). *Ecotourism and sustainable development: Who owns paradise?* Island Press.

Hossen, A., Ahsan, M. F., & Kamruzzaman, M. (2015). Ecotourism potentiality in the Dulahazra Safari Park, Bangladesh. *Journal of Environmental Science and Natural Resources, 7*, 189–196.

Kavallinis, I., & Pizam, A. (1994). The environmental impacts of tourism—Whose responsibility is it anyway? The case study of Mykonos. *Journal of Travel Research, 33*, 26–32.

Meneghello, S., & Mingotto, E. (2016). Promoting sustainable development through fisheries-related tourism experiences. Benefits from the integration between fisheries and tourism in venetian coastal areas. *International Journal of Sustainable Development and Planning, 11*, 447–457.

Sarker, S., Bhuyan, M. A., Rahman, M. M., Islam, M. A., Hossain, M. S., Basak, S. C., & Islam, M. M. (2018). From science to action: Exploring the potentials of blue economy for enhancing economic sustainability in Bangladesh. *Ocean and Coastal Management, 157*, 180–192.

Togiak National Wildlife Refuge Core Planning Team. (2001). *Togiak National Wildlife Refuge: Comprehensive conservation plan revision: Significant planning issues*. Togiak National Wildlife Refuge.

Zambrano, A. M., Broadbent, E. N., & Durham, W. H. (2010). Social and environmental effects of ecotourism in the Osa peninsula of Costa Rica: The Lapa Rios case. *Journal of Ecotourism, 9*, 62–83.

Role of Sport Fisheries in Conservation 22

Abstract

Recreational fishing is an important part of the world's aquatic ecosystems; it provides ecological, social, and economic advantages but also presents obstacles to the preservation of biodiversity. This chapter highlights the significance of sustainable management approaches by examining the dual effects of recreational fishing on fish populations and habitats. Angler stewardship and economic contributions are important assets, but there are still vulnerabilities in the form of hazards from invasive species and gaps in our understanding of ecology. Opportunities exist for reducing threats from overfishing and habitat degradation, as well as for community engagement and conservation education. The integration of scientific research and local knowledge is emphasized in the guiding principles for sustainable development, which aim to promote mutually beneficial outcomes for conservation and fishery initiatives.

Keywords

Recreational fisheries · Biodiversity conservation · Sustainable management · Angler stewardship · Habitat degradation

1 Introduction

Recreational fisheries encompass activities that capture and possibly harvest aquatic animals for reasons other than meeting primary physiological (i.e., nutritional) needs, where the catch is usually not traded on formal markets or otherwise sold (EIFAC, 2008; Arlinghaus & Cooke, 2009). Popular worldwide and pursued in both marine and inland waters, recreational fishing is typically performed with rod and line using natural or artificial baits, although other methods are employed according to local customs (Arlinghaus & Cooke, 2009). In temperate zones, recreational

© The Author(s), under exclusive license to Springer Nature Singapore Pte Ltd. 2024

S. T. Mushtaq et al., *Angling in India*,
https://doi.org/10.1007/978-981-97-8876-7_22

fisheries represent an important use of fish resources (Cowx & Arlinghaus, 2008; Smith, 1986; Welcomme, 2001).

With increasing industrialization, the significance of commercial fishing decreases, and inland water bodies are increasingly or exclusively utilized for recreational fisheries. This trend persists, with management objectives increasingly focusing on non-fishery-related conservation goals when anthropogenic changes become evident within a society (Arlinghaus & Cooke, 2009). Globally, approximately 10.6% of the population in the industrialized world participates in recreational fisheries. Similarly, many transitional economies in Africa, Asia, and Latin America are shifting from small-scale commercial and subsistence fisheries towards recreational fisheries, especially as economic development fosters leisure time and tourism (Ditton, 2008; Ellender et al., 2010; Potts et al., 2009).

Despite the considerable economic potential of recreational fisheries, it is often unrecognized or underestimated by political decision-makers (Arlinghaus, 2006a; Cooke & Cowx, 2004, 2006; Cowx, 1998; Hickley, 2009). Furthermore, there are concerns that activities associated with recreational fishing conflict with non-fishery-related aquatic biodiversity conservation objectives, such as those advocated under the Millennium Ecosystem Assessment (2005) and various conventions and directives enacted through the Convention for Biodiversity (CBD). Conflicts often arise from divergent perspectives on the intrinsic versus instrumental values of biodiversity in the context of conservation biology and fisheries, particularly when recreational fishing exploits already threatened fish resources and aquatic ecosystems.

Many threats to aquatic biodiversity originate from external sources, such as habitat degradation, pollution, nutrient inputs, and damming, rather than the recreational fisheries sector itself (Arlinghaus et al., 2002; Cowx, 2002). Effective decision-making in conservation must build on the instrumental value of biodiversity to engage stakeholders (Justus et al., 2009). For this article, nature conservation in the context of aquatic biodiversity and recreational fisheries is defined as the conservation of the structure and diversity of all biological levels of an aquatic ecosystem, including its surrounding terrestrial ecosystem, in a state that existed without recreational fisheries exploitation and management. This perspective values larger biological entities, such as populations, equally with supporting habitats and individuals making up these populations (Paterson, 2006).

2 SWOT Analysis

A SWOT analysis identifies and analyzes the strengths and weaknesses of a sector, in this case, recreational fisheries, as well as the opportunities and threats revealed by the external environment (externalities). This analysis helps develop a plan that considers various internal (strengths and weaknesses) and external (opportunities and threats) factors, maximizing the potential of strengths and opportunities while minimizing the effects of weaknesses and threats (Fig. 22.1).

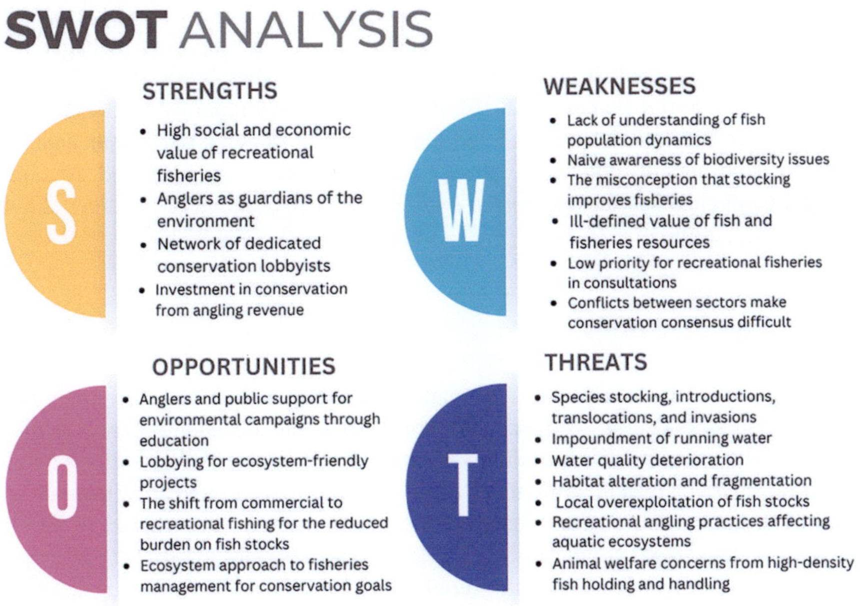

Fig. 22.1 SWOT analysis of recreational fisheries

2.1 Strengths

Economic, social, and ecological benefits generated through recreational fisheries have been comprehensively reviewed by Weithman (1999). These benefits encompass classical economic advantages to angler-dependent industries, social welfare to anglers, and other social and psychological benefits. Recognizing these benefits is crucial, as they financially contribute to conservation efforts and foster widespread aquatic stewardship (Knuth & Siemer, 2007).

Anglers have a vested interest in preserving or enhancing the resources they depend on. Evidence suggests that anglers proactively work to conserve and enhance aquatic biodiversity, either directly through supportive stocking of native fishes or indirectly through habitat management and other fisheries management actions, usually financed by angling license money (Granek et al., 2008).

2.2 Weaknesses

A significant weakness is the occasional lack of understanding of the ecological processes driving fish population dynamics among some angler communities and fisheries managers, particularly under private fishing rights regimes where managers are usually self-educated, avid anglers (Arlinghaus, 2006b). This misconception fosters the belief that fisheries can only be improved by intensive stocking (Cowx,

1999) and contributes to unrealistic expectations among anglers about the natural limits of fisheries (North, 2002). Additionally, there is pressure on managers by local anglers to establish and translocate non-native fishes, some of which may become invasive (Gozlan et al., 2010). As a result, recreational fisheries managers in some countries consistently manage stocks through stocking rather than enhancing the environment, potentially disrupting ecosystem functioning and affecting several ecological services generated by fish stocks (Holmlund & Hammer, 2004; Lewin et al., 2008).

2.3 Opportunities

Successful conservation projects often involve people and incorporate their attitudes, wishes, concerns, beliefs, and preferences into decision-making (Meffe, 2002). The general public, including anglers, can be valuable ambassadors for conservation causes, given their numbers, dispersion, and vested interest in maintaining healthy fisheries and environments (Granek et al., 2008). The challenge lies in increasing the general populace's and anglers' awareness of the complex issues facing aquatic biodiversity. Promoting environmental education and extension programs can harness their willingness to support environmental and conservation campaigns (Knuth & Siemer, 2007).

2.4 Threats

Aquatic biodiversity is threatened by various factors, primarily stemming from anthropogenic disturbances, including species stocking, introductions, translocations, invasions, impoundment of running water, water quality deterioration, habitat alteration and fragmentation, and overexploitation (Richter et al., 1997). These threats are universal, and although environmental legislation addresses many issues in developed countries, progress is slow. Water resource development schemes pose particular problems as their economic value often outweighs the perceived fisheries, biodiversity, and conservation values of aquatic ecosystems. The cost of implementing rehabilitation programs or finding alternative solutions to water resource demands is often prohibitive, leading to either a status quo or progressive deterioration of habitat quality.

Recreational fisheries could play a key role in reversing these problems, but given the low social priority of inland fisheries, the outlook is not positive. Nonetheless, the recreational fisheries sector is a powerful lobby that demands good environmental quality to maintain fisheries, complementing conservation sector aspirations (Kearney, 2002). It is critical for conservation bodies to engage with the fisheries sector to present unified arguments for remedial actions benefiting both fisheries and aquatic biodiversity.

3 Impacts on Individual Fish Welfare

Recreational fishing also affects the welfare and well-being of individual fish not harvested. Concerns about the holding of fish at high density in keepnets or enclosures, along with the hooking, playing, and handling of captured fish, potentially causing injury and distress, have prompted numerous studies on catch-and-release mortality and sublethal alterations (e.g., physiology and behavior) in recreational fisheries (Bartholomew & Bohnsack, 2005). Generally, studies suggest that if fish are appropriately handled and injury and stress are minor, recovery is rapid with no long-term effects (Cooke et al., 2007). Additionally, studies on holding fish in keepnets indicate minimal stress until high densities are reached or inappropriate facilities are used (Pottinger, 1997). In most countries, anglers are aware of animal welfare issues and strive to minimize the impact on fisheries and wildlife (Berg & Rosch, 1998). However, animal rights campaigners increasingly challenge recreational fishing, particularly catch-and-release, questioning the humaneness of angling (Huntingford et al., 2006).

4 Guidelines for Developing Best Practices

Developing sportfishing in developing countries requires the consideration of complex cultural, social, and environmental factors. The following guidelines can help achieve sustainable sportfishing outcomes:

1. Develop a comprehensive understanding for the full spectrum of values (historical and cultural) of fisheries to local communities.
2. Understand, respect, and comply with local governance (formal and informal) related to ownership and the use of natural resources and environments where sport fishing occurs.
3. Inform, consult, and involve local communities in decisions related to sport fishing development.
4. If necessary, provide location-specific guidelines for culturally appropriate behavior to sport fishing clients and operators.
5. Engage with local leaders to establish and support continuous and efficient lines of communication with local stakeholders (e.g., technical working groups or advisory committees).

5 Managing the Resource and Environment

Sport fishing ventures into pristine and isolated areas, presenting opportunities and challenges for conservation and resource management. The appeal of catching unique species in wild areas drives sport fishing into increasingly remote locations, impacting unique species and habitats. Effective management requires a deep understanding of the biology, ecology, and conservation status of targeted species, informing sustainable fishing practices and minimizing ecological impacts (FAO, 2012). Developing sport fisheries in developing countries necessitates robust scientific research and monitoring to ensure optimal outcomes. By integrating conservation goals with local community needs and cultural sensitivities, sport fishing can contribute positively to livelihoods and environmental stewardship.

6 Options for the Future

One of the major challenges in the discourse between recreational fisheries and conservation is the widespread lack of ecological knowledge about aquatic species abundance, distribution, and population dynamics. This issue highlights the need for better governance and management practices, focusing on sustainable stock-enhancement strategies, optimal regulatory policies, and best catch-and-release practices.

Continually developing and implementing best practice guidelines tailored to individual cases can help align sportfishing tourism with ecological and social criteria. Such guidelines, based on site-specific research, have the potential to provide mutual benefits to tourists and local communities, promoting environmental stewardship and enhancing cultural experiences. By adopting a sustainable livelihood framework, sportfishing development can enhance local livelihoods and contribute to environmental conservation. This integrated approach ensures that sportfishing initiatives are both ecologically and socially sustainable, benefiting both local communities and the broader environment.

7 Conclusion

The intersection of recreational fisheries and aquatic biodiversity conservation presents a complex but promising landscape. While recreational fisheries generate significant economic, social, and ecological benefits, there are notable weaknesses and threats, such as habitat degradation, invasive species, and the potential for overfishing. However, with appropriate management and increased ecological awareness among anglers, these challenges can be mitigated. Effective conservation strategies must incorporate stakeholder involvement, community engagement, and the development of best practices tailored to local contexts. By fostering a collaborative approach between recreational fishers and conservationists, it is possible to enhance both fish populations and their habitats, ensuring sustainable and mutually beneficial outcomes. The chapter underscores the necessity of integrating recreational fisheries management with broader conservation objectives to protect and sustain aquatic biodiversity for future generations.

References

Arlinghaus, R., T. Mehner, and I. G. Cowx. (2002). Reconciling traditional inland fisheries management and sustainability in industrialized countries, with emphasis on Europe. Fish and Fisheries 3:261–316.

Arlinghaus, R. (2006a). *Der untersch¨atzte Angler*. Kosmos.

Arlinghaus, R. (2006b). Overcoming human obstacles to conservation of recreational fishery resources, with emphasis on Central Europe. *Environmental Conservation, 33*, 46–59.

Arlinghaus, R., & Cooke, S. J. (2009). Recreational fisheries: Socioeconomic importance, conservation issues and management challenges. In B. Dickson, J. Hutton, & W. A. Adams (Eds.), *Recreational hunting, conservation and rural livelihoods: Science and practice* (pp. 39–58). Blackwell Publishing.

Bartholomew, A., & Bohnsack, J. A. (2005). A review of catch-and-release angling mortality with implications for no-take reserves. *Reviews in Fish Biology and Fisheries, 15*, 129–154.

Berg, R., & Rosch, R. (1998). Animal welfare and angling in Baden-Wurttemberg, Germany. In P. Hickley & H. Tompkins (Eds.), *Recreational Fisheries: Social, economic and management aspects* (pp. 88–92). Fishing News Books, Blackwell Science.

Cooke, S. J., & Cowx, I. G. (2004). The role of recreational fishing in global fish crises. *Bioscience, 54*, 857–859.

Cooke, S. J., & Cowx, I. G. (2006). Contrasting recreational and commercial fishing: Searching for common issues to promote unified conservation of fisheries resources and aquatic environments. *Biological Conservation, 128*, 93–108.

Cooke, S. J., Suski, C. D., Ostrand, K. G., Philipp, D. P., & Wahl, D. H. (2007). Physiological and behavioral consequences of long-term artificial selection for vulnerability to recreational angling in a teleost fish. *Physiological and Biochemical Zoology, 80*, 480–490.

Cowx, I. G. (1998). Aquatic resource management planning for resolution of fisheries management issues. In P. Hickley & H. Tompkins (Eds.), *Recreational fisheries: Social, economic and management aspects* (pp. 97–105). Fishing News Books, Blackwell Science.

Cowx, I. G. (1999). An appraisal of stocking strategies in the light of developing country constraints. *Fisheries Management and Ecology, 6*, 21–34.

Cowx, I. G. (2002). Analysis of threats to freshwater fish conservation: Past and present challenges. In M. J. Collares-Pereira, I. G. Cowx, & M. M. Coelho (Eds.), *Conservation of freshwater fish: Options for the future* (pp. 201–220). Blackwell Science.

Cowx, I. G., & Arlinghaus, R. (2008). Recreational fisheries in the twenty-first century: Towards a code of conduct. In Ø. Aas, R. Arlinghaus, R. B. Ditton, D. Policansky, & H. L. Schramm Jr. (Eds.), *Global challenges in Recreational Fisheries* (pp. 338–352). Blackwell Science.

Ditton, R. B. (2008). An international perspective on recreational fishing. In Ø. Aas, R. Arlinghaus, R. B. Ditton, D. Policansky, & H. L. Schramm Jr. (Eds.), *Global challenges in recreational fisheries* (pp. 5–55). Blackwell Science.

Ellender, B. R., Weyl, O. L. F., Winker, H., Stelzhammer, H., & Traas, G. R. L. (2010). Estimating angling effort and participation in a multi-user, inland fishery in South Africa. *Fisheries Management and Ecology, 17*, 19–27.

FAO. (2012). Recreational Fisheries. In *FAO Technical Guidelines for Responsible Fisheries. No. 13* (p. 176). FAO.

Gozlan, R. E., Britton, J. R., Cowx, I. G., & Copp, G. H. (2010). Current knowledge on nonnative freshwater fish introductions. *Journal of Fish Biology, 76*, 751–786.

Granek, E. F., Madin, E. M. P., Brown, M. A., Figueira, W., Cameron, D. S., Hogan, Z., Kristianson, G., de Villiers, P., Williams, J. E., Post, J., Zahn, S., & Arlinghaus, R. (2008). Engaging recreational fishers in management and conservation: Global case studies. *Conservation Biology, 22*, 1125–1134.

Hickley, P. (2009). Recreational fishers—Social economic and management aspects. In *Fisheries, sustainability and development* (pp. 169–188). Royal SwedishAcademy of Agriculture and Forestry.

Holmlund, C. M., & Hammer, M. (2004). Effects of stocking on ecosystem services: An overview and case study using the Stockholm archipelago. *Environmental Management, 33*, 799–820.

Huntingford, F. A., Adams, C., Braithwaite, V. A., Kadri, S., Pottinger, T. G., Sandøe, P., & Turnbull, J. F. (2006). Current issues in fish welfare. *Journal of Fish Biology, 68*, 332–372.

Justus, J., Colyvan, M., Regan, H., & Maguire, L. (2009). Buying into conservation: Intrinsic versus instrumental value. *Trends in Ecology & Evolution, 24*, 187–191.

Kearney, R. E. (2002). Recreational fishing: Value is in the eye of the beholder. In T. J. Pitcher & C. E. Hollingworth (Eds.), *Recreational fisheries: Ecological, economic and social evaluation* (pp. 17–33). Blackwell Publishing.

Knuth, B. A., & Siemer, W. F. (2007). *Aquatic stewardship education in theory and practice* (p. 55). American Fisheries Society Symposium.

Lewin, W.-C., McPhee, D., & Arlinghaus, R. (2008). Biological impacts of recreational fishing resulting from exploitation, stocking and introduction. In Ø. Aas, R. Arlinghaus, R. B. Ditton, D. Policansky, & H. L. Schramm Jr. (Eds.), *Global challenges in Recreational Fisheries* (pp. 75–92). Blackwell Science.

Meffe, G. K. (2002). Connecting science to management and policy in freshwater fish conservation. In M. J. Collares-Pereira, I. G. Cowx, & M. M. Coelho (Eds.), *Conservation of freshwater fish: Options for the future* (pp. 363–372). Fishing News Books, Blackwell Science.

North, R. (2002). Factors affecting the performance of Stillwater coarse fisheries in England and Wales. In I. G. Cowx (Ed.), *Management and ecology of Lake and reservoir Fisheries* (pp. 284–298). Blackwell Science.

Paterson, B. (2006). Ethics for wildlife conservation: Overcoming the human-nature dualism. *Bioscience, 56*, 144–150.

Pottinger, T. G. (1997). Changes in water quality within anglers' keepnets during confinement of fish. *Fisheries Management and Ecology, 4*, 341–354.

Potts, W. M., Childs, A.-R., Sauer, W. H. H., & Duarte, A. D. C. (2009). Characteristics and economic contribution of a developing recreational fishery in southern Angola. *Fisheries Management and Ecology, 16*, 14–20.

Richter, B. D., Braun, D. P., Mendelson, M. A., & Master, L. L. (1997). Threats to imperiled freshwater fauna. *Conservation Biology, 11*, 1081–1093.

Smith, C. L. (1986). The life cycle of fisheries. *Fisheries, 11*, 20–25.

Weithman, A. S. (1999). Socioeconomic benefits of fisheries. In C. C. Kohler & D. D. Hubert (Eds.), *Inland fisheries management in North America* (2nd ed., pp. 193–213). American Fisheries Society.

Welcomme, R. L. (2001). *Inland Fisheries: Ecology and management*. Blackwell Science, Fishing News Books.

State Regulations Governing Sport Fisheries

23

Abstract

Anglers from all over the world are drawn to India by the wide range of game fish species that are supported by the country's varied inland and marine habitats. However, a number of anthropogenic factors pose a threat to these precious resources' long-term viability. State governments in India have put in place extensive laws governing the management and conservation of sport fisheries to guarantee the continuation of the tradition of enjoying sport fishing. These rules usually cover things like granting licenses and permits for fishing, creating sanctuaries and protected areas, imposing size and catch limitations, and encouraging catch-and-release methods. States have different restrictions that are tailored to the particular features of their game fish species and local ecosystems. The goal of state governments' enforcement of these laws is to achieve a balance between the recreational and economic benefits of sport fishing, and the ecological preservation of these iconic game fish species. Effective implementation and enforcement of these regulations, coupled with collaborative efforts involving anglers, researchers, and conservation organizations, are crucial for the long-term sustainability of India's sport fisheries.

Keywords

Catch and release · Licenses · Catch limits · Protected areas · Marine habitats

1 Introduction

Sport fishing (recreational or game fishing) is widely regarded as one of the most thrilling recreational pursuits worldwide. Many developed and developing nations boast advanced tools and techniques for catching fish. However, the pursuit of this pastime is not without its responsibilities. Regulations governing sport fishing serve

© The Author(s), under exclusive license to Springer Nature Singapore Pte Ltd. 2024

S. T. Mushtaq et al., *Angling in India*,
https://doi.org/10.1007/978-981-97-8876-7_23

as guardians of aquatic ecosystems, ensuring the sustainable utilization of fishery resources while preserving the natural balance of our waters. However, in countries like India, there exists a significant gap in information and scientific understanding regarding sport fishing, primarily due to its region- and species-specific nature. While some Indian states adhere to government regulations, there is a lack of awareness about the need to conserve threatened and endangered fish species through sport fishing activities. Furthermore, promoting this activity could address unemployment by generating income and fostering ecotourism opportunities. Government intervention is essential for the sustainable development and management of sport fisheries in India (Mandal et al., 2018). Fishery laws and regulations are a framework established to safeguard fish populations in public waters. They are designed to promote conservation efforts and prevent harmful fishing practices. Throughout history, there has been a cultural awareness of the importance of preserving natural fishery resources, as evidenced by ancient practices in India. According to the Food and Agriculture Organization (FAO, 2009), fisheries management encompasses a comprehensive process involving information gathering, analysis, planning, consultation, decision-making, resource allocation, and the formulation and implementation of regulations to ensure sustainable resource utilization and achieve various fisheries objectives. The production and yield of fisheries are influenced by diverse factors such as biological, ecological, environmental, technological, social, cultural, and economic considerations. Common management strategies include regulating mesh size of nets, imposing size limits on fish, setting catch quotas, restricting fishing methods and times, issuing licenses and leases for fishing rights, establishing sanctuaries, and prohibiting certain fishing structures like anicuts and fixed engines (Edwin, 2022).

2 Global Impact of Recreational Fishing

Recreational fishing has made significant contributions to both regional and national economies globally, as highlighted by studies conducted by Cooke and Cowx (2004), Cowx et al. (2010), Danylchuk and Cooke (2011), and Everard and Kataria (2008). However, it is important to note that recreational fishing has also been linked to adverse effects on biodiversity and ecosystems, as indicated by research from Lewin et al. (2006). Collaborative efforts between recreational fishers and local stakeholders have resulted in conservation successes, particularly in initiatives aimed at protecting threatened and endangered species. Studies by Arlinghaus et al. (2002), Arlinghaus (2006), Granek et al. (2008), and Cowx et al. (2010) have highlighted the positive outcomes of such collaborations. However, addressing the species-specific impacts of recreational fisheries requires research and management investments. Unfortunately, in India, there exists a gap between policy makers and anglers, hindering such investments. This emphasizes the importance of gaining a deeper understanding of the sector to guide its responsible development (Pinder & Raghavan, 2013).

To address harmful fishing practices, the Indian Fisheries Act was enacted in 1897 with the primary goal of regulating such methods (Mandal et al., 2018). In addition to Indian Fisheries Act, various regulations govern the protection of certain species. Some fish species may be designated as protected under State or Central Wildlife protection laws. These protected species are generally prohibited from being caught or retained. If accidentally caught, they must be released immediately to preserve their populations and ensure their long-term survival. Fishing seasons and catch limits are other essential components of sport fishing regulations in India. Many states implement specific fishing seasons to protect fish during their breeding or spawning periods, allowing them to reproduce and sustain their populations. Catch limits are also established to ensure that anglers do not excessively exploit fish stocks, maintaining a sustainable balance between recreational fishing and conservation.

3 Regional Regulations and Conservation Efforts

Different states have specific regulations governing recreational fishing to ensure sustainability. These regulations aim to strike a balance between recreational fishing opportunities and the preservation of fish populations and aquatic ecosystems. By adhering to these regulations, anglers can enjoy their favorite pastime while contributing to the sustainable management and conservation of India's diverse fish species.

Himachal Pradesh, Jammu and Kashmir, Uttarakhand, Assam, and Maharashtra each have unique regulations tailored to their ecosystems and angling opportunities. These regulations emphasize responsible angling practices, seasonal restrictions, and penalties for destructive methods.

3.1 Himachal Pradesh

In Himachal Pradesh, sport fishing is governed by specific regulations to ensure sustainable practices and the preservation of aquatic ecosystems. With over 3000 km of rivers and streams, the state offers diverse opportunities for anglers. Of this vast network, approximately 600 km are designated for trout fishing, while the remaining areas cater to species like mahseer and schizothoracids. The key rivers include Beas, Sutlej, Ravi, Tirthan, Sainj, Uhl, Baspa, Pabar, Lambadug, Cir, Rana, Nugal Gai, Baner, and Bata. Licenses are issued for fishing sites along these rivers, with regulations stipulating the use of rod and line only, in accordance with the Himachal Pradesh Fisheries Act. The Department of Fisheries has identified specific stretches totaling about 80 km for trout fishing, primarily on rivers like Beas, Tirthan, Sainj, Uhl, Ravi, and Lambadug. Similarly, designated areas spanning 5–10 km along the Beas River are earmarked for mahseer fishing, totaling over 50 km. Well-known fishing spots include upstream of Rohru in the Pabbar Valley, Uhl river near Barot, and upper reaches of the Beas and its tributaries. To facilitate angling activities, the Directorate of Fisheries provides information on license issuing authorities, fees, reservation of rest houses, and contact details of relevant officers. Additionally, guidelines are provided to anglers, including seasonal bans to protect spawning periods. Fishing is prohibited in all general waters, including lakes and reservoirs, during the spawning season from June first to July 31st and in trout waters from November first to February 28th of the following year. The regulations also emphasize the prohibition of destructive practices such as dynamite fishing and the use of poisonous substances. Offenders may face severe penalties, including imprisonment or fines. Himachal Pradesh, particularly the Kullu Manali region and Larji Valley, has emerged as a prime destination for anglers, boasting streams rich in brown trout. The Himachal Pradesh government has introduced a unique provision within the Fisheries Act, stipulating that fishing during the breeding season constitutes a serious offense, punishable by imprisonment of upto 3 years, and classified as cognizable and non-bailable.

3.2 Jammu and Kashmir

Jammu and Kashmir offers approximately 1050 km of rivers, renowned for their potential as trout streams, including the Bringhi, Sindh, Aru, Kishanganga, Lidder, Boniyar, Dagwan, and Erin rivers. Additionally, elevated lakes such as Vishansar, Kishansar, Gangabal, and others above 3500 m are prime fishing areas, abundant with brown trout. Permits for trout fishing are issued by the Department of Fisheries in Srinagar, with the best fishing season spanning from April to October. Snow-trout fishing is feasible across the valley, particularly along the Jhelum River and its tributaries. Notable sites include the Arh stream feeding Dal Lake, Madhumati feeding Wular Lake, and Mirhuma feeding the Jhelum River. While mahseer fishing isn't prevalent in the valley, it thrives in the Jammu region, particularly in rivers like Chenab and Ravi. Fishing sites such as Ikhnee Nalla and Anji offer opportunities for anglers seeking mahseer. The Fisheries Act of Jammu and Kashmir, 2018 likely contains regulations regarding sport fishing, which include guidelines for licenses, permissible catch limits and designated areas. Penalties for offences could range from fines to potential revocation of fishing privileges, depending on severity of violation. Regulations governing sport fishing in Jammu and Kashmir prioritize responsible angling practices to preserve the region's natural beauty and aquatic ecosystems. Fishing permits are necessary for designated areas, with strict adherence to seasonal regulations to protect spawning periods and ensure sustainable fishing practices. Anglers have the option to secure angling licenses/permits conveniently through the department website (Fig. 23.1). The fee for angling is levied on a daily basis, with foreign anglers charged Rs. 2000/day and local anglers charged Rs. 1000/day.

Organized brown trout fishing is primarily limited to the streams of Kashmir and Himachal Pradesh in the northwestern Himalayas (Fig. 23.2). Allowed baits for brown trout fishing include "dry and wet" fly spinning, artificial and natural worms, etc. Artificial spinning bait has been prohibited in Kashmir waters since 1970 (Sehgal, 1987). In Himachal Pradesh, all methods of sport fishing are permissible

Department of Fisheries, J&K
Trout Angling Bookings Management System
Booking Confirmation Slip

Booking ID :	657		
Booking Date :	29/4/2024		
Angler's Name & Parentage	DR MUSADIQ ALAQABAND S/O MUSHTAQ AHMED ALAQABAND		
Mobile No :	7006048515	Email ID :	musadiq200@gmail.com
R/O :	Bemina, Srinagar		
Nationality :	INDIAN	Country :	India
Passport No / National ID Card No (for foreigners) :			

Your Booking for the following beat has been confirmed.

Date of Angling	2/5/2024	Beat Name	Phrysian Upper
Stream	Lidder	District	ANANTNAG

Amount deposited : Rs.1000/- vide Transaction no : 4400170211977

This booking is valid only for the date of angling **(ONE DAY ONLY)**. and **1 Rod only**

Note:

1. This permit should be handed over to the fisheries guard of the beat.
2. Bag limit is six fish of above 25cm size.
3. use of spinning rod, spinners, or spinning reel, live bait are strictly prohibited.
4. Service of registered ghilee (shikari) is essential.
5. Esteemed anglers are advised to report to the beat by 07.00AM for better results.
6. Fishing tackle should be of good quality.
7. The permit is non-refundable, non transferable, and no change in respect of beat / date is acceptable, whatsoever will be the reasons, including bad road, bad weather etc.
8. For assistance, please contact 0194-2500092, 2500106, 2500124

By Order
Director of Fisheries,
Govt of J&K

Fig. 23.1 Angling permit provided by Department of Fisheries, J&K

Fig. 23.2 Brown trout from Kashmir beats. Picture credits: Dr. Musadiq Allaqaband

under the bylaws. Trout streams in Kashmir are divided into fishing beats, each with a stream length of 3–5 km. The number of anglers allowed in each beat is regulated on a daily, weekly, or seasonal basis. The fishing season extends from March to October every year. The minimum legal size for trout caught in Himalayan streams in Kashmir and Himachal Pradesh ranges from 25–30 cm. The bag limit ranges from 5–7 fish of 25 cm and above in length. Anglers must retain undersized fish caught in Kashmir but return them in Himachal Pradesh, recording such catches in a logbook. However, compliance with these instructions is limited among anglers (Magray, 2016).

3.3 Uttarakhand

Uttarakhand's pristine rivers, streams, and lakes offer abundant opportunities for fishing enthusiasts. The Ramganga and Sharda rivers, along with Dodital Lake, are renowned for trout and mahseer fishing. Additionally, stretches around Begs Ghat and Gangalehri on the Ganga and Yamuna rivers, as well as their tributaries in the Dhoon valley, provide excellent mahseer fishing grounds. While brown trout populations are limited in rivers like Bhairathi, Birehi, and Pinder, efforts are underway to enhance trout rearing in the region. The Directorate of Coldwater Fisheries Research (DCFR) operates the Chirapani Farm in Champawat district, rearing rainbow trout to table size. The state also boasts four hatcheries, including the Bairanga Trout Farm and Chirapani Farm, producing fingerlings of both brown and rainbow trout. Hatcheries at Talwari and Kaldyani focus on rainbow and brown trout respectively, albeit facing challenges related to water quality, disease, and feed. Plans are in place to expand trout aquaculture to support stocking rivers, lakes, and raceways for sport fishing. Additionally, a mahseer hatchery at Bhimtal has been established by the DCFR, producing golden mahseer seed since 1990. These seeds contribute to stocking various lakes in the Kumaon Himalayas, including the Kosi River, and supply other agencies. Regulations governing sport fishing in Uttarakhand prioritize sustainability and conservation efforts. Anglers are encouraged to obtain necessary permits and adhere to guidelines to ensure the preservation of aquatic ecosystems while enjoying the abundant fishing opportunities Uttarakhand has to offer.

Angling spots along the rivers of Uttarakhand are designated into beats. These beats are leased to local Self-Help Groups (SHGs), Wildlife Help Groups (WHGs), Youth Help Groups (YHGs), etc., for the purpose of conducting angling activities. Angling permits are issued to both national and international tourists for engaging in angling activities within the designated beats. The rates for these permits vary based on the zone and the type of tourist. The Government of Uttarakhand has prescribed the following angling permit rates (Table 23.1):

3.4 Assam

Angling holds a prominent place among the favorite sports in Assam, with the mighty Brahmaputra River, its tributaries, and numerous smaller streams offering an array of fishing sites. Specifically, the Jia Bhoroli, Kapili, and Manas rivers are renowned for their excellent mahseer (Tor spp.) and katli (*N. hexagonolepis*) fishing

Table 23.1 Angling permit rates prescribed by Uttarakhand Govt

	Mahseer zone	Trout zone
National tourist	Rs. 75 per permit	Rs. 100 per permit
International tourist	Rs. 150 per permit	Rs. 200 per permit
Local people (for employment)	Rs. 20 per permit	Rs. 20 per permit

grounds. Each November, an annual competition draws hundreds of anglers from both within and outside the country to participate in fishing at the Jia Bharali River, aiming to land the prized golden mahseer. Notably, the record catch of a golden mahseer at Jia Bharali on rod and line weighed an impressive 24.5 kg. Recognizing the rapid decline in the population of golden mahseer due to poaching, the Assam (Bhoreli) Angling and Conservation Association has taken proactive measures. These include stocking the river with fingerlings and implementing scientific breeding programs to safeguard the species. These conservation efforts aim to ensure the preservation of the rich angling heritage of Assam for generations to come.

3.5 Maharashtra

Powai Lake, an artificial reservoir created in 1891 to supply drinking water to Mumbai, has transformed into a haven for anglers over the past century, earning its reputation as an angler's paradise in the city. Initially leased to the Western India Fishing Association, the Bombay Presidency Angling Association took over in 1936. In 1955, it was renamed the Maharashtra State Angling Association (MSAA), focusing on angling as a sport while also addressing the preservation of Powai Lake due to its deteriorating environment. The MSAA has played a vital role in stocking the lake with various species including catla, rohu, mrigal, calbasu, khudree mahseer, and others. However, pollution, poaching, water hyacinth infestation, and silting have adversely impacted the water quality and aquatic life in the lake. To counteract these challenges, the association has implemented measures such as fish rearing and release programs from its nursery, aimed at replenishing losses due to poaching. Despite the environmental challenges, Powai Lake remains a focal point for angling enthusiasts. The MSAA has invested in covered boats, a club house, and ongoing efforts to conserve the lake's ecosystem. Furthermore, the association's efforts have led to Powai Lake's inclusion in the National Lake Conservation Programme of the Government of India, highlighting its significance in the conservation of aquatic resources. These regulations underscore the importance of sustainable angling practices and environmental conservation in ensuring the continued enjoyment of Powai Lake for generations to come.

3.6 Sea Sport Fishing

Sea sport fishing hasn't flourished in India, yet the country boasts a diverse range of game fish such as Barracuda, Mullet, Perch, Tuna, Marlin, Sailfish, and Seerfish along its coasts. Since 1993, the government has permitted "Sea Game Fishing," particularly around the Andaman Islands. Renowned as one of the premier game fish destinations globally, the Andaman & Nicobar Islands attract enthusiasts. Additionally, select coastlines like Odisha, Lakshadweep, Goa, and Kerala offer setups for open sea game fishing. Chilka Lake in Odisha provides opportunities to fish for Tuna and Prawn, while the waters surrounding Lakshadweep and Andamans are known for Tuna, Blue Marlin, Barracuda, and Grouper.

3.7 Odisha

Located on India's east coast, Odisha boasts unique game fishing opportunities in Chilka, its largest brackish water lake. Abundant with Tiger Prawn, Tuna, and various other fish species, Chilka Lake is also renowned for its migratory bird life.

3.8 Lakshadweep

The Lakshadweep Islands, nestled in the Arabian Sea, are teeming with game fish such as sharks, Kulvel, Barracuda, blue marlin, and groupers. Sport fishing is permitted only on Agatti and Bangaram Islands, using the pole and line method to target big fish. The fishing season spans from October to May.

3.9 Andaman Islands

Comprising around 293 islands stretching across the Bay of Bengal, the Andaman Islands stand as one of the world's premier game fish destinations. Commercial fishing by long-lines and trawlers is prohibited, with traditional hand line fishing practiced by local inhabitants. Sea sport fishing is permitted around the south Andaman Islands between October and April, with Game Fish Charters allowed from February to April. A catch and release policy is adopted to ensure sustainable angling.

3.10 Goa and Kerala

The sea fishing region spanning from Goa to Thiruvananthapuram along the southwest coast offers prime angling sites. Notable ports include Goa, Mangalore, Cochin, and Thiruvananthapuram, with top fishing spots along the Konkan coast including Alibag, Harihareshwar, Kankavli, and Vijaydurg. Optimal fishing times coincide with high tide, utilizing sardines or shrimp as bait.

4 Angling Associations of India: National and Regional Networks

Sport fishing regulations for both freshwater and marine species are primarily governed by the Indian Fisheries Act (1897). Additionally, local angler associations have established specific rules to protect their interests and address ecological considerations. In India, a network of both national and regional angling organizations supports the sport fishing industry, with each entity committed to its promotion and regulation. These associations play pivotal roles in nurturing a community of anglers, championing conservation efforts, and upholding the sustainability of fish populations. Nationally, numerous organizations cater to anglers nationwide, while regionally, diverse groups focus on specific states or territories, tailoring their initiatives to the distinct ecosystems and angling opportunities within their respective areas (Table 23.2). Collaborating closely with governmental bodies, conservation organizations, and local communities, these regional associations strive to enforce regulations, conduct research, and raise awareness about the importance of sustainable angling practices. Through collective efforts, both national and regional angling associations contribute significantly to safeguarding India's varied aquatic ecosystems and ensuring the continued enjoyment of sport fishing for generations to come.

Table 23.2 Various organizations associated with sport fishing in India

Target Region		Organization	Target species
National	Throughout India	All India game fish association (AIGFA)	All
		International game fish association (IGFA)	All
		Indian angler	All
Regional	Assam	Assam angling and conservation association	Mahseer
	Arunachal Pradesh	Kemang angling association	Mahseer, trout
		Pasighat angling club	Mahseer, trout
	Andaman	Game fishing India	All
	Himachal Pradesh	Trout conservation & angling association	Trout
	Kerala	Cochin anglers	Marine fish
		High range angling association	Trout
	Karnataka	Wildlife Association of South India (WASI)	Mahseer
		Coorg wildlife society (CWS)	Mahseer
	Maharashtra	Maharashtra state angling association (MSSA)	Carp
		Anglers association, Fulata Lake	Carp
		Naushad Ali Sarovar Samvardhani (NASS)	Mahseer, trout
	Nagaland	Nagaland anglers association	Mahseer, trout
	Sikkim	Sikkim anglers association	Mahseer, trout
	Tripura	Tripura angling association	Mahseer, trout
	Tamil Nadu	Chennai anglers association	Marine fish

Source: Gupta et al., 2015

5 Conclusion

The regulations governing sport fishing in India reflect the rich mosaics of historical significance, conservation efforts, and regional initiatives aimed at preserving aquatic ecosystems and promoting sustainable angling practices. From ancient cultural practices to modern-day conservation efforts, India's approach to fisheries management underscores the importance of balancing recreational enjoyment with environmental stewardship. Regulatory measures aimed at conserving fish populations have evolved over time, reflecting a deep-rooted understanding of the need to protect aquatic resources. In recent decades, collaborative efforts between governmental agencies, non-profit organizations, and angling associations have led to significant strides in conservation and regulation. Initiatives such as catch-and-release angling, habitat restoration, and species-specific conservation programs have contributed to the preservation of iconic fish species like mahseer and trout. State-specific regulations tailored to unique ecosystems and angling opportunities further highlight India's commitment to responsible fisheries management. From the Himalayan streams of Himachal Pradesh to the brackish waters of Odisha's Chilka Lake, regulations prioritize sustainability while ensuring recreational opportunities for anglers. Furthermore, the establishment of national and regional angling organizations underscores the collective effort to promote sport fishing while advocating for conservation principles. These organizations serve as catalysts for community engagement, education, and advocacy, fostering a culture of responsible angling across the country. As India continues to navigate the complexities of fisheries management in the face of environmental challenges and evolving recreational trends, the importance of collaborative stewardship cannot be overstated. By embracing science-based regulations, community participation, and a shared commitment to conservation, India is poised to ensure the enduring legacy of sport fishing for generations to come.

References

Arlinghaus, R. (2006). Overcoming human obstacles to conservation of recreational fishery resources, with emphasis on Central Europe. *Environmental Conservation, 33*, 46–59.

Arlinghaus, R., Mehner, T., & Cowx, I. G. (2002). Reconciling traditional inland fisheries management and sustainability in industrialized countries, with emphasis on Europe. *Fish and Fisheries, 3*, 261–316.

Cooke, S. J., & Cowx, I. G. (2004). The role of recreational fishing in global fish crises. *Bioscience, 54*, 857–859.

Cowx, I. G., Arlinghaus, R., & Cooke, S. J. (2010). Harmonizing recreational fisheries and conservation objectives for aquatic biodiversity in inland waters. *Journal of Fish Biology, 76*, 2194–2215.

Danylchuk, A. J., & Cooke, S. J. (2011). Engaging the recreational angling community to implement and manage aquatic protected areas. *Conservation Biology, 25*, 458–464.

Edwin, L. 2022. *Training manual: ICAR-sponsored short course on bycatch reduction in fisheries: Recent advances*

Everard, M., & Kataria, G. (2008). Recreational angling markets to advance the conservation of a reach of the Western Ramganga River, India. *Aquatic Conservation: Marine and Freshwater Ecosystems, 21*, 101–108.

FAO. (2009). *A fishery manager's guidebook. Management measures and their application. FAO Fisheries Technical Paper No. 424* (p. 231). Rome, FAO.

Granek, E. F., Madin, E. M. P., Brown, M. A., Figueira, W., Cameron, D. S., Hogan, Z., Kristianson, G., de Villiers, P., Williams, J. E., Post, J., Zahn, S., & Arlinghaus, R. (2008). Engaging recreational fishers in management and conservation: Global case studies. *Conservation Biology, 22*, 1125–1134.

Gupta, N., Bower, S. D., Raghavan, R., Danylchuk, A. J., & Cooke, S. J. (2015). Status of recreational fisheries in India: Development, issues, and opportunities. *Reviews in Fisheries Science & Aquaculture, 23*(3), 291–301.

Lewin, W. C., Arlinghaus, R., & Mehner, T. (2006). Documented and potential biological impacts of recreational fishing: Insights for management and conservation. *Reviews in Fisheries Science, 14*, 305–367.

Magray, N. A. (2016). Sport Fishing, an Opportunity for Promotion of Adventure Tourism in Jammu and Kashmir. *International Journal in Management & Social Science, 4*(11), 112–123.

Mandal, A., Das, S. K., Biswas, B., & Kairnar, S. O. (2018). Future scope of sport fisheries development in India: a review. *Indian Journal of Animal Health, 57*(1), 1–16.

Pinder, A. C., & Raghavan, R. (2013). Conserving the endangered Mahseers (Tor spp.) of India: The positive role of recreational fisheries. *Current Science, 104*, 1472–1475.

Sehgal, K. L. (1987). *Sport fisheries of India* (p. 126). Indian Council of Agricultural Research.

Future Scope of Sport Fishing in India 24

Abstract

Sport fishing in India has a bright future because of the nation's varied aquatic environments and expanding recreational fishing market, which draw anglers from all over the world. There is a great chance to advance and promote sustainable sport fishing techniques given the rising popularity of game fish species including Mahseer, Trout, and marine fishes. Future priorities should include growing fishing tourists, putting strong conservation measures in place, and improving data gathering and study on game fish stocks. India can secure the long-term survival of its famous fish species and establish itself as a top worldwide sport fishing destination by investing in infrastructure, hiring guides, and controlling fishing activity. Achieving this goal and realizing the full potential of sport fishing in India would require cooperation between public and private sector organizations as well as fishing enthusiasts.

Keywords

Aquatic environments · Angler · Fish stocks · Guides · Fishing tourists

1 Introduction

Recreational angling, a popular leisure activity worldwide, is gaining traction in India. Historically, this sport has been male dominated, shaped by societal norms that often see sports as a male pursuit. However, there is a growing recognition of the need to diversify and expand participation. Angling associations are pivotal in this transformation, working to create opportunities for women and promote an inclusive environment. By doing so, not only can female participation be boosted, but local communities can also benefit economically. The involvement of "angling families" can stimulate cottage industries and local businesses, fostering economic

© The Author(s), under exclusive license to Springer Nature Singapore Pte Ltd. 2024

S. T. Mushtaq et al., *Angling in India*,
https://doi.org/10.1007/978-981-97-8876-7_24

growth and social development (Gupta et al., 2015a, b, c). The evolving landscape of recreational angling in India also highlights the importance of engaging younger generations, addressing governance challenges, fostering collaboration for effective management, and adopting sustainable practices. The recreational fisheries sector in India holds significant promise, yet several pressing issues must be addressed to ensure its sustainable development.

1.1 Lack of Best Practices for Sustainable Angling

Currently, there are no official guidelines or monitoring systems for recreational fisheries in India. Angling associations often self-regulate, promoting best management practices during events like the "All India Fresh Water Angling Competition" (Gupta et al., 2015a, b, c). However, these practices are not uniformly adopted, leading to harmful behaviors such as the use of improper hooks and inadequate handling of fish. Increasing awareness and encouraging compliance through education and enforcement are essential for promoting sustainable angling practices.

1.2 Unregulated Stocking and Introduction of Species

The development of recreational fisheries in India has been partly supported by the introduction of both exotic and captive-bred native species. Historical stocking of salmonids in upland lakes and rivers during the British Raj has impacted endemic species (Sehgal, 1999). The large-scale introduction of various mahseer species in the Cauvery River has led to hybridization and potential declines in native populations (Dinesh et al., 2010). There is a need for careful assessment of these practices to understand their ecological impacts and ensure the genetic integrity of native fish populations.

1.3 Governance Challenges

The multi-jurisdictional nature of fisheries governance in India complicates decision-making at both national and state levels (Raghavan et al., 2012). The absence of a centralized governing body with legislative support hinders the effective management and regulation of recreational fisheries. Additionally, the disconnect between recreational fisheries associations and government agencies creates obstacles for sustainable development.

1.4 Insufficient Knowledge on Game Fish

Research on freshwater fish in India is notably sparse, with limited data on their biology, ecology, and population dynamics (Dahanukar et al., 2011). This lack of information extends even to well-known species like the mahseer. Uncertainties persist regarding the exact number of mahseer species and their specific distribution in India (Pinder & Raghavan, 2013). For instance, the "Cauvery humpbacked mahseer," a popular target for anglers, still awaits scientific classification despite its popularity.

1.5 Understanding Biotic Responses to Catch and Release

There is a significant gap in research on the impact of catch-and-release (C&R) practices on fish species in India. No studies have examined post-capture mortality rates for mahseer or other commonly targeted species, leading to assumptions that many fish may die due to exhaustion, injuries, or infections (Dinesh et al., 2010). The type of fishing gear used can influence mortality rates, with natural baits often causing deeper hooking and higher mortality compared to artificial lures (Cooke & Schramm, 2007). Additionally, the use of barbless and circle hooks has been shown to reduce handling time and tissue damage, promoting lower mortality rates (Cooke et al., 2001).

1.6 Stakeholder Engagement and Conflicts

Interactions with anglers have highlighted a lack of government support and enforcement against illegal fishing practices. There is also a need for better integration of local communities in the management of recreational fisheries to ensure they benefit from the industry. Establishing freshwater fish safe zones monitored by local communities could address conflicts between stakeholders and support both recreational and subsistence fishing (Gupta et al., 2015a, b, c).

1.7 Lack of Representative Data

One of the major challenges is the absence of comprehensive data on recreational fisheries in India. Collecting representative data is difficult due to the variability in accessibility across different regions. Innovative sampling methods, such as face-to-face interviews or angler diaries, may be necessary to obtain accurate data. Collaboration between government agencies, fishing organizations, and local communities will be crucial for funding and conducting these surveys (Gupta et al., 2015a, b, c).

Addressing these issues through targeted research, improved governance, stakeholder engagement, and sustainable practices is essential for the development and conservation of the recreational angling sector in India. In order to improve the status of the sector some management approaches are practised in India which include:

1.7.1 Licensing and Fees

Requiring recreational anglers to purchase fishing licenses and pay associated fees helps manage participation levels and provides funding for conservation efforts.

1.7.2 Gear Restrictions

Limiting the types of fishing gear and methods allowed helps prevent damage to target fish species and their habitats.

1.7.3 Method Restrictions

Restricting certain fishing methods, such as banning destructive fishing in certain areas, reduces damage to species and habitats.

1.7.4 Closed Times and Seasons

Prohibiting fishing during spawning and migration periods helps create a less stressful environment for fish.

1.7.5 Closed Areas

Establishing protected areas that prohibit fishing helps safeguard spawning grounds and migration routes.

1.7.6 Bag and Size Limits

Restricting the number and size of fish that can be retained helps prevent overharvesting of desirable species.

1.7.7 Sale Prohibitions

Banning the commercial sale of recreationally caught fish discourages the commercialization of recreational fisheries.

1.7.8 Harvest Restrictions

Prohibiting the targeting of threatened or endangered species helps protect vulnerable populations.

1.7.9 Fish Holding Restrictions

Regulations against translocating or holding live fish help minimize stress and damage to species.

These types of regulations are crucial for managing recreational fisheries in a sustainable manner by controlling participation, protecting habitats and species, and preventing overharvesting.

2 Future Prospects

2.1 Engaging the Younger Generation

In October 2012, the AIGFA co-organized a day-long angler's camp for children aged seven to nine at the WASI lakes in Karnataka. Participants were educated about the different fish species in the lake, the importance of catch and release angling for environmental conservation, and the basics of angling equipment and its assembly (Gupta et al., 2015a, b, c).

2.2 Resolving Governance Challenges

There is an urgent need to address the governance issues surrounding freshwater fisheries management in India, particularly concerning angling locations. Some parts of water bodies fall within protected areas (PAs) and are thus under the jurisdiction of state forest departments. However, forest managers often claim control over the entire water body, leading to disputes with local stakeholders. Protecting the entire stretch of a water body is crucial for the survival of species within the PA, as external anthropogenic stressors can have devastating impacts (Gupta et al., 2014a, b; 2015a, b, c).

2.3 Collaboration for Effective Management

To effectively manage freshwater ecosystems and species, there needs to be a consensus between state fisheries departments, local stakeholders, and forest managers. Recognizing village communities as key stakeholders in conservation plans can enhance the success of recreational fisheries. Associations should integrate these communities into their long-term strategies (Gupta et al., 2014a, b).

2.4 Sustainable Stocking Practices

It is essential to understand that fish stocking should be conducted under specific conditions to avoid diminishing genetic diversity. The IUCN Guidelines for Reintroductions and other Conservation Translocations (IUCN, 2012) emphasize that reintroductions should benefit both the species and its ecosystem and be based on thorough scientific research. Therefore, stocking merely to increase catch size should be avoided, particularly as the genetic makeup of many target fish populations, including mahseer, remains unknown.

2.5 Advancing Recreational Fisheries Management

India's recreational fisheries management approaches must be refined to provide long-term ecological, social, and economic benefits and support additional freshwater conservation projects (Gozlan et al., 2013; Rogers, 2013).

3 Conclusion

Sport fishing in India has a bright future because of the nation's varied aquatic environments and expanding recreational fishing market, which draw anglers from all over the world. India can establish itself as one of the world's top sport fishing destinations by boosting research and data collection, developing fishing tourism, and investing in sustainable practices. The potential of this beloved leisure activity may be fully realized by cooperative initiatives combining government agencies, the private sector, and fishing enthusiasts, while also guaranteeing the long-term survival of India's unique game fish species.

References

Cooke, S. J., & Schramm, H. L. (2007). Catch-and-release science and its application to conservation and management of recreational fisheries. *Fisheries Management and Ecology, 14*, 73–79.

Cooke, S. J., Philipp, D. P., Dunmall, K. M., & Schreer, J. F. (2001). The influence of terminal tackle on injury, handling time, and cardiac disturbance of rock bass. *North American Journal of Fisheries Management, 21*, 333–342.

Dahanukar, N., R. Raghavan, A. Ali, R. Abraham, And C. P. Shaji (2011) The status and distribution of freshwater fishes of the Western Ghats, p. 116. In: The status and distribution of freshwater biodiversity in the Western Ghats, India. (Molur, S., K. G. Smith, B. A. Daniel, and W. R. T. Darwall. : IUCN/Zoo Outreach Organization.

Dinesh, K., Nandeesha, M. C., Nautiyal, P., & Aiyappa, P. (2010). Mahseers in India: A review with focus on conservation and management. *The Indian Journal of Animal Sciences, 80*, 26–38.

Gozlan, R. E., Burnard, D., Andreou, D., & Britton, J. R. (2013). Understanding the threats posed by non-native species: Public vs. conservation managers. *PLoS One, 8*, e53200.

Gupta, N., Sivakumar, K., Mathur, V. B., & Chadwick, M. A. (2014a). The "tiger of Indian rivers": Stakeholders' perspectives on the golden mahseer as a flagship fish species. *Area, 46*, 389–397.

Gupta, N., Raghavan, R., Sivakumar, K., & Mathur, V. B. (2014b). Freshwater fish safe zones (FFSZs): A prospective conservation strategy for river ecosystem in India. *Current Science, 107*, 949–950.

Gupta, N., Sivakumar, K., Mathur, V. B., & Chadwick, M. A. (2015a). Terrestrial protected areas and managed reaches conserve threatened freshwater fish in Uttarakhand, India. *PARKS, 21*(1), 89–101.

Gupta, N., Raghavan, R., Sivakumar, K., Mathur, V. B., & Pinder, A. C. (2015b). Assessing recreational fisheries in an emerging economy: Knowledge, perceptions and attitudes of catch-and-release anglers in India. *Fisheries Research, 165*, 79–84.

Gupta, N., Bower, S. D., Raghavan, R., Danylchuk, A. J., & Cooke, S. J. (2015c). Status of recreational fisheries in India: Development issues, and opportunities. *Reviews in Fisheries Science & Aquaculture, 23*(3), 291–230.

Pinder, A. C., & Raghavan, R. (2013). Conserving the endangered Mahseers (tor spp.) of India: The positive role of recreational fisheries. *Current Science, 104*, 1472–1475.

Raghavan, R., Ali, A., Dahanukar, N., & Rosser, A. (2012). Is the Deccan mahseer, tor khudree (Sykes 1839) (Pisces: Cyprinidae) fishery in the Western Ghats hotspot sustainable? A participatory approach to stock assessment. *Fisheries Research, 110*, 29–38.

Rogers, A. A. (2013). Social welfare and marine reserves: Is willingness to pay for conservation dependent on management process? A discrete choice experiment of the Ningaloo Marine Park in Australia. *Canadian Journal of Agricultural Economics, 61*, 217–238.

Sehgal, K. L. (1999). Coldwater fish and fisheries in the Indian Himalayas: Rivers and streams. In *Fish and fisheries at higher altitudes: Asia. FAO fisheries technical paper. No. 385*. FAO.

Glossary

Abrasion Resistance Abrasion is the process of scuffing, scratching, wearing down, marring, or rubbing away.

Acceleration Measured as external acceleration (EA), which is the recorded three-axis acceleration above standard gravity defined with static calibration, normalized, and calculated as the vectorial sum in milli-g.

Acidosis A condition caused by excess acid in the body fluids.

Acrobatic Fish behavior, especially in species known for jumping or leaping out of the water, such as trout or salmon during spawning runs.

Aerodynamics The study of the properties of moving air and the interaction with solid objects.

Alpine Lake A lake situated at a high altitude, typically in mountainous areas.

Angling Permits Licenses required for fishing.

Angling The sport or pastime of fishing with a rod and line.

Anicuts Small dams or weirs built across streams or rivers to manage water flow.

Anthropogenic Resulting from the influence of human beings on nature.

Arbor Knot The Arbor knot is a typical fishers' knot used to attach fishing line to the arbor of a fishing reel.

Archipelago A group or chain of islands.

Artificial Lures Man-made objects designed to attract fish.

Artificial Propagation Breeding and raising fish in controlled environments.

Back Cast A short backward and often upward swing of a fishing rod, its line, and its lure in preparation for the cast that immediately follows.

Bag Limit The maximum number of fish that can be legally caught in a day.

Bait Casting A method of fishing where the bait is cast using a baitcasting reel.

Bait Throwing Fishing bait is any luring substance used specifically to attract and catch fish, typically when angling with a hook and line.

Bait Fishing bait is any luring substance used specifically to attract and catch fish, typically when angling with a hook and line.

Barbed Having barbs or sharp projections, often used to describe fishing hooks.

Barbels Whisker-like sensory organs near the mouths of some fish species.

Barbless Hook Hooks without barbs, designed to minimize injury to fish.

© The Author(s), under exclusive license to Springer Nature Singapore Pte Ltd. 2024
S. T. Mushtaq et al., *Angling in India*,
https://doi.org/10.1007/978-981-97-8876-7

Barotrauma Injury caused by changes in pressure, affecting fish brought up from deep water.

Bass Fishing Bass fishing is the recreational fishing activity, typically via rod-based angling, for various game fishes of North America known collectively as black bass.

Beads Small, often colorful, spheres used in fishing rigs for various purposes.

Beats Designated areas for fishing.

Beats Designated areas for fishing.

Benthopelagic Feeder Organisms that feed both near the bottom and in the water column.

Biodiversity Hotspots Areas that are rich in biodiversity and threatened by human activities.

Biodiversity The variety of all forms of life on earth, including species diversity, genetic diversity, and ecosystem diversity.

Boat Fishing Fishing from a boat.

Boat Rigs Fishing rigs used from a boat.

Boat A vessel used for fishing or transportation on water.

Bobber A float used in fishing to suspend bait at a certain depth.

Bottom Bouncing A fishing technique where the bait is bounced along the bottom.

Bottom Feeders A bottom feeder is an aquatic animal that feeds on or near the bottom of a body of water.

Bottom Fishing Fishing near the bottom of the water column.

Branchiostegites The bony plates that cover and protect the gills of fish.

Breaking Strength Breaking strength or Fracture strength is the stress at which a specimen fails via fracture.

Brine It may refer to the salt solutions ranging from about 3.5% (a typical concentration of seawater, on the lower end of that of solutions used for brining foods) up to about 26% (a typical saturated solution, depending on temperature).

Buoyancy The ability to float in water.

Buoyancy It is a gravitational force, a net upward force exerted by a fluid that opposes the weight of a partially or fully immersed object.

Bycatch Non-target species that are unintentionally caught while fishing.

Cane Pool Traditional fishing pole made from cane.

Canoe A lightweight, narrow boat pointed at both ends and open on top, propelled by one or more seated paddlers facing the direction of travel using single-bladed paddles.

Canoeing The sport or activity of paddling a canoe.

Carolina Rig A fishing rig used for presenting soft plastic baits near the bottom.

Cast Jig A casting jig is another extremely versatile style of jig.

Cast Netting A method of fishing using a circular net that is thrown by hand.

Casting Stroke When the rod applies force to the line in order to form a loop, the casting stroke ends with loop formation.

Casting the Line Throwing the fishing line out into the water.

Casting The act of throwing the fishing line into the water.

Cataract A large, powerful waterfall or a medical condition affecting the eye.

Catch and Release Policy A fishing practice where caught fish are released back into the water.

Catch and Release A practice in which fish are caught and then returned to the water.

Catch Composition The variety and quantity of species caught in fishing.

Catch Limits Restrictions on the number or size of fish that can be caught.

Catch Quotas Limits on the amount or number of fish that can be caught.

Cephalic Pits Sensory pits on the head of some fishes, used to detect vibrations.

Chunk A thick, solid piece that has been cut or broken off something.

Circle Hook A type of hook designed to catch fish in the corner of the mouth.

Clinch Knot A knot used to secure a fishing line to a hook or lure.

Clipped Down Fin clipping is the most common method for marking groups of fish, but there is still debate in the scientific community as to whether it causes harm.

Clippers A clipper is a type of fast sailing vessel, generally from the nineteenth century.

Closed Fishing Periods when fishing is not allowed to protect fish populations.

Cognizable and Non-Bailable Legal terms referring to offenses that can be recognized by a police officer without a warrant and for which bail is not available.

Commercial Fishery Fishing operations that catch fish for sale and profit.

Conservation The protection and preservation of the natural environment and wildlife.

Coral Reef Underwater structures made of coral, important habitats for marine life.

Crank Bait A type of lure designed to mimic the swimming action of fish.

Crank A crank is an arm attached at a right angle to a rotating shaft by which circular motion is imparted to or received from the shaft.

Current The continuous flow of water in a particular direction.

Daisy Chains A series of lures or baits attached together.

Deep Sea Fishing Fishing in the open ocean, typically far from land.

Defrost A defrosting procedure is generally performed periodically on refrigerators and freezers to maintain their operating efficiency.

Destructive Fishing Fishing practices that harm ecosystems and deplete fish populations.

Detecting Fish Bites Recognizing when a fish is biting the bait.

Disoriented Confused about place or direction.

Docks A dock is an enclosed area in a harbor where ships go to be loaded, unloaded, and repaired.

Dodgers Flashing devices used to attract fish.

Drag System The drag is simply a pair of friction plates inside of fishing reels. If the fish pulls on the line hard enough, the friction is overcome, and the reel rotates backwards, letting line out, preventing the line from breaking.

Drop Shutting Gradually reducing or completely stopping water flow to a fishing area, often for maintenance or management purposes.

Dropper Loops Knots used in fishing to create a loop for attaching hooks or lures.

Dynamite Fishing Using explosives to stun or kill fish for easy collection, which is highly destructive to marine environments.

Ecology The study of interactions among organisms and their environment.
Ecosystem A community of living organisms and their physical environment interacting as a system.
Ecotourism Ecotourism is one such strategy which has often been promoted by integrated conservation and development projects (ICDPs).
Elastomer An elastomer is a polymer with viscoelasticity (i.e., both viscosity and elasticity) and with weak intermolecular forces, generally low Young's modulus (E) and high failure strain compared with other materials.
Emarginated Having a notch or indentation at the edge or tip, often used to describe the shape of fins.
Endangered Species Species at serious risk of extinction.
Endemic Species Species that are native to and found only in a specific geographic area.
Endemic Species Species that are native to and found only in a specific geographic area.
Estuaries Coastal water bodies where freshwater from rivers and streams meets and mixes with saltwater from the ocean.
Eversion Turning inside out, such as a fish's stomach due to pressure changes.
Exhaustion Extreme physical or mental fatigue.
Exotic Trouts Trout species that are not native to a particular area.
Exotic Not native, introduced from another area.
Expedition A journey or voyage undertaken for a specific purpose, often related to exploration or research.
Exploitation Utilization of resources for human benefit, often referring to overuse.
False Casting It is periodically keeping that line up in the air when one is either drying a dry fly, changing direction, gauging distance, or gauging accuracy.
Feverish Pulsation Pulse fishing is a fisheries management technique for preventing fish stocks from being overfished by periodically permitting a cycle of fishing followed by a fallow period which allows stocks to reconstitute.
Finesse Jigging A fishing technique using lightweight jigs and subtle presentations.
Fingerling A young fish that is about the size of a finger.
Finlets Small fins located behind the dorsal and anal fins of some fish species.
Fire Extinguisher A device used to put out fires.
Fish Rearing The process of raising fish, typically in a controlled environment.
Fish Safe Zone Areas designated to protect fish populations.
Fishing Charters Services that take people on guided fishing trips.
Fishing Cooperatives Groups of fishers working together for mutual benefit.
Fishing Crafts Boats and other equipment used for fishing.
Fishing Permits Licenses required to fish in certain areas.
Fishing Season Specific times of the year when fishing is allowed to ensure sustainability.
Fishing Spool The part of the fishing reel that holds the line.
Fixed Engines Permanent fishing devices, such as traps or nets, fixed in one place.
Fizzing or Venting The process of releasing gas from a fish's swim bladder.

Flies An artificial fly or fly lure is a type of fishing lure, usually used in the sport of fly fishing (although they may also be used in other forms of angling).

Flipping Flipping involves peeling off about 50–75% or more line than the length of the rod and simply feeding the line back through the guides as you drop and lift the rod.

Floats Buoyant devices used in fishing to suspend bait at a desired depth.

Fly Casting A method of casting using a lightweight fly line.

Fly Fishing Fly fishing is an angling technique that uses an ultra-lightweight lure called an artificial fly, which typically mimics small invertebrates such as flying and aquatic insects to attract and catch fish.

Fly Tying It is the process of producing an artificial fly used by fly fishing anglers to catch fish.

Forages The term "forage fish" is a term used in fisheries, and is applied also to forage species that are not true fish but play a significant role as prey for predators.

Forward Cast When you fly fish, you cast forwards and backwards in a back-and-forth motion.

Free Lining Fishing without any added weight or float.

Friction Friction is the force resisting the movement of solid surfaces and materials sliding against each other.

Frogging Fishing for frogs.

Front Drag The drag is simply a pair of friction plates inside of fishing reels.

Game Fishing Fishing for sport rather than for food or commercial purposes.

Game Wild animals hunted for sport or food.

Gill Arch A bony or cartilaginous structure on each side of the pharynx that supports the gills.

Gill Nets Fishing nets designed to catch fish by their gills.

Gill Rakers Bony or cartilaginous projections which point forward and inward from the gill arches, involved in filter feeding.

Glacial Rivers Rivers originating from glaciers.

GPS Service A system that provides location and time information.

Ground Baiting The practice of throwing bait into the water to attract fish.

Guidelines A set of principles or rules to follow.

Gular Plate A plate located in the throat region of some fishes and amphibians.

Habitat Degradation The process by which a habitat becomes incapable of supporting its native species.

Habitat Destruction The complete elimination of a habitat, making it impossible for species to survive there.

Habitats The natural environments in which organisms live.

Harappa Another significant archaeological site in the Indus Valley civilization.

Hatcheries Facilities where fish are bred, hatched, and reared.

Heat Stroke A severe heat-related illness.

Hematoma A localized collection of blood outside the blood vessels.

Hemorrhaging Bleeding, especially severe or uncontrolled bleeding.

Hemostats Clamping tools used to remove hooks from fish.

Heritage Cultural or natural inheritance.

High-Low A fishing rig with two hooks at different heights.
Hook Removers Tools designed to safely remove hooks from fish.
Hook Shanks The straight part of a hook.
Hooks Curved devices used to catch fish by hooking into their mouth.
Hooks Devices for catching fish.
Hydrographic Basin A land area where all precipitation collects and drains off into a common outlet, such as a river, bay, or other body of water.
Hypoxia A deficiency of oxygen.
Indian Peninsula The region of South Asia extending into the Indian Ocean.
Indus Valley Sites Locations of ancient settlements along the Indus River, known for early civilization.
Infection The invasion and multiplication of microorganisms in body tissues.
Infestation The presence of a large number of unwanted organisms in a place.
Injuries Physical harm or damage to the body.
Interorbital Area The region between the eyes of a fish or other animal.
Island Hopping Traveling from one island to another.
Jetties Structures extended into the water to protect a harbor or influence currents.
Jig Head A type of fishing lure with a weighted head.
Jig A jig consists of a heavy metal (typically lead) sinker with an attached fish hook that is usually obscured inside a soft lure or feather-like decorations.
Jigging Jigging is the practice of fishing with a jig, a type of weighted fishing lure.
Jigs A jig is a versatile tool that attracts small, medium and large fish.
Jurisdiction The official power to make legal decisions and judgments, often relating to specific geographic areas.
Kayak A small, narrow watercraft propelled by a double-bladed paddle.
Kayaking The activity of paddling a kayak, a small, narrow watercraft which is typically propelled by means of a double-bladed paddle.
Kite Used in kite fishing to suspend baits on the surface of the water.
Knocker Rig A fishing rig used to present bait close to the bottom.
Land Slides The movement of rock, earth, or debris down a slope.
Leader A length of line attached to the end of the main fishing line, often stronger or more invisible.
Leases Contracts granting use or occupation of property during a specified period in exchange for rent.
Ledger A method of fishing where the bait is kept on the bottom.
License Official permit to engage in fishing activities.
Life Jacket A flotation device worn to keep a person afloat in water.
Line Chumming Using a line with bait attached to attract fish.
Line The cord or thread used in fishing to attach the hook and bait.
Log Books Records kept by fishers documenting their catches and other relevant information.
Long Lining A commercial fishing technique using long lines with baited hooks.
Lubrication Lubrication is the process or technique of using a lubricant to reduce friction and wear and tear in a contact between two surfaces.

Lures A fishing lure is any one of a broad category of artificial angling baits that are inedible replicas designed to mimic prey animals (e.g., baitfish, crustaceans, insects, worms, etc.) that attract the attention of predatory fish, typically via appearances, flashy colors, bright reflections, movements, vibrations and/or loud noises which appeal to the fish's predation instinct and entice it into gulping the lure.

Mangrove Coastal ecosystems characterized by salt-tolerant trees and shrubs.

Marine Protected Area A region of the ocean designated for conservation and restricted use.

Maritime Waters Waters that are part of or adjacent to the sea.

Maxilla The upper jawbone in vertebrates.

Mechanized Operated by machinery.

Mega Biodiversity Centers Regions with a high level of biodiversity.

Mesh Size The size of the openings in a fishing net, which determines what size fish are caught.

Microhabitat A small, specific habitat within a larger ecosystem.

Migration The seasonal movement of fish for spawning or feeding.

Minnow Minnow is the common name for a number of species of small freshwater fish, belonging to several genera of the family Cyprinidae and in particular the subfamily Leuciscinae.

Mohenjo-daro An archaeological site in the Indus Valley, known for its ancient urban planning.

Monitoring Observing and recording fishing activities to ensure compliance.

Monofilament Monofilament is made by melting and mixing polymers and then extruding the mixture through tiny holes, forming strands of line, which is then spun into spools of various thicknesses.

Monofilament Monofilament is made by melting and mixing polymers and then extruding the mixture through tiny holes, forming strands of line, which is then spun into spools of various thicknesses.

Fluorocarbon line Fluorocarbon fishing line is made of the fluoropolymer PVDF and it is valued for its refractive index, which is similar to that of water, making it less visible to fish.

Monsoon A seasonal wind pattern that brings heavy rains.

Mortality Death, especially on a large scale.

Mud Water Water with a high concentration of suspended sediment.

National Sanctuary Protected areas established to conserve wildlife and natural resources.

Native Species Species that occur naturally in a specific area.

Native Indigenous or original to a particular region or environment.

Needle Nosed Pliers Pliers with long, slender jaws used in fishing and other tasks.

Nightcrawlers Nightcrawler, any large earthworm, especially those favored in angling.

Normal Fee The standard charge for a service or activity.

Nursing Nurturing or rearing fish.

Nymphs Aquatic insect larvae used as fishing bait.

Occipital Process A projection at the back of the skull.

Ocellus An eye-like spot found on some animals, often used for camouflage or signaling.

Open Sea Game Fishing Fishing in the open ocean, typically for large sport fish.

Operculum A bony flap that covers and protects the gills in fishes.

Over Exploitation Excessive use of resources, leading to depletion.

Overfishing Harvesting fish to the point where species are depleted and the balance of the ecosystem is disturbed.

Overharvesting The excessive removal of fish from a body of water, leading to population decline.

Palomar Knot A strong and simple knot used to attach a fishing line to a hook or lure.

Peg An area set aside for an angler

Penalties Punishments imposed for breaking laws or regulations.

Pickerel Wire A type of wire used in making fishing rigs.

Plankton Small and microscopic organisms drifting or floating in water.

Pliers Fishing pliers are used to remove hooks from fish.

Plugs Saltwater fishing plugs are one of the most common types of lures used by both inshore and offshore anglers.

Poaching Illegal fishing or hunting.

Poisonous Substances Chemicals used to catch fish that can also harm other marine life and ecosystems.

Pole and Line A traditional method of fishing using a rod (pole) and baited hook (line).

Pollution The presence of harmful substances in the environment.

Poppers Fishing lures designed to make a popping sound on the water's surface.

Pound Test Rating The measurement of the strength of fishing line.

Primitive Relating to an early stage in evolutionary or historical development.

Prolapse The falling down or slipping of a body part.

Propeller A device with blades that rotate to propel a boat or lure.

Radar Reflectors Devices used to improve a vessel's visibility on radar.

Rain Forest Dense, tropical forests with high annual rainfall.

Recreational Fishing Techniques include hand gathering, spearfishing, netting, angling and trapping.

Recreational Activities done for enjoyment and relaxation.

Recreational Activities done for enjoyment and relaxation.

Reel A device used to wind and store fishing line.

Reels Devices used to wind and store fishing line.

Regulation Rules or directives made and maintained by an authority.

Reintroduction The process of returning animals to their native habitats.

Reservoirs Artificial lakes used for water storage.

Rig A rig is an arrangement of items used for fishing.

Rigging Live Baits Rigging and gear setup are critical elements in successful live bait fishing, particularly when targeting large and powerful marine predators.

River Jigging Fishing in rivers using jigs.

Rivers Large, natural streams of water flowing in channels to the sea, a lake, or another river.

Rod The long, flexible pole used in fishing to cast the line and reel in fish.

Roe Fish eggs used as bait or food.

Rugose Having a rough, wrinkled surface.

Salt Water Surf Fishing from the shore or wading in the surf in saltwater environments.

Sanctuaries Protected areas where fishing and other potentially harmful activities are restricted.

Scuba Diving Underwater diving using a self-contained underwater breathing apparatus (scuba).

Sea Game Fishing Fishing for sport in the open sea.

Seasonal Bans Temporary prohibitions on fishing during certain times of the year to protect fish populations.

Seasons and Area Regulations specifying when and where fishing is allowed.

Serpentine These are greenish, brownish, or spotted minerals commonly found in serpentinite.

Shaft The main body or handle of a fishing rod.

Shank The straight part of a fish hook.

Shanks The straight part of a fish hook.

Shooting line A running/shooting line is a crucial part of your Spey fishing set-up.

Shore The land along the edge of a sea, lake, or other large body of water.

Shoreline The edge of a body of water.

Sinkers Weights used in fishing to sink the bait or lure.

Sinkers Weights used in fishing to sink the bait or lure.

Sinkers Weights used to hold the bait or lure at the desired depth.

Slipping Slipping is a somewhat unique case of discarding, in that the unwanted catch (or component that is unwanted) is released whilst still in the water and before being brought aboard the fishing vessel.

Slitting Cutting fish to remove the guts.

Snagging Catching fish by hooking them anywhere on the body rather than the mouth.

Snap Swivel A combination of a snap and a swivel, used to quickly change lures or rigs.

Snappers The snapper, *Pagrus auratus* (Bloch and Schneider) (Sparidae) is another important recreational and commercial food fish that occurs broadly in Australian waters, as well as in many other areas of the Indo-Pacific in the northern and southern hemispheres.

Snoods Short lengths of line attached to the main fishing line with hooks on the end.

Snorkeling Swimming with a mask and snorkel to view underwater life.

Soldered Joined or repaired using solder.

Spawning Grounds Specific areas where fish go to reproduce.

Spawning Migration The movement of fish from one location to another to reproduce.

Spawning Periods Times of the year when fish reproduce.

Spawning Season The spawn is another name for the mating process carried out by fish.
Species Specific Pertaining to or characteristic of a particular species.
Spin Casting A fishing technique using a spin casting reel.
Spinner Bait A type of lure with one or more spinning blades.
Spinners Lures with spinning blades to create flash and vibration.
Spinners Spin fishing is an angling technique where a spinner bait, a type of hybrid fishing lure with at least one freely rotating blade, is used to entice the fish to bite.
Spooking Scaring fish away.
Spool Fishing line, thread, and wire are usually wound around a special cylinder called a spool.
Spools The spool is the part of a reel that holds the fishing line, but it also plays an important role in casting distance and smoothness.
Spoon A type of fishing lure that imitates small fish by wobbling through the water.
Spreader Bars Devices used to spread multiple lures out to prevent tangling.
Stillwaters Stillwater fishing (or still fishing) is a matter of putting your bait in the water and waiting for a fish to find it, making it one of the simplest and most effective fishing methods for beginners.
Stocks Populations of fish available for fishing.
Streamer A type of fly fishing lure that imitates baitfish.
Stress Physical or emotional tension.
Stressors Factors that cause stress to organisms, affecting their health and survival.
Striking Setting the hook in a fish's mouth by a sharp pull on the line.
Stroke Zone The area in the water where fish are likely to strike.
Submerged Covered or under water.
Subsistence Fishing for personal or family consumption rather than commercial purposes.
Surf Casting Casting a fishing line from the shore into the surf.
Surf Fishing Fishing from the shore or wading in the surf.
Surfing The sport of riding waves on a surfboard.
Surveillance Keeping watch over fishing areas to prevent illegal activities.
Sustainability The ability to maintain ecological and resource balance over time.
Sustainable Exploit Using resources in a way that does not deplete them for future generations.
Sweeping Moving a fishing rod in a sweeping motion to set the hook.
Swell Section Part of the water with large waves.
Swimbaits Swimbaits or swimmers are a loosely defined class of fishing lures that are designed to primarily imitate the underwater swimming motions of baitfishes.
Swivel Usually made of metal, a fishing swivel connects to the pivoting joint with a rod and reel that are tied to one end.
Swivels Small devices placed between the fishing line and the leader to prevent line twisting.
Swivels Small devices placed between the fishing line and the leader to prevent line twisting.
Tackle Equipment used in fishing, such as rods, lines, hooks, and lures.

Tackle Fishing tackle is the equipment used by anglers when fishing.

Tarpons Are considered some of the greatest saltwater game fishes, prized not only because of their great size but also because of their fight and spectacular leaping ability.

Tensile Strength Ultimate tensile strength is the maximum stress that a material can withstand while being stretched or pulled before breaking.

Terminal Tackle attached to the end of a fishing line that gets cast out along with the bait are referred to as terminal tackle.

The Texas Rig It is a technique used for fishing soft plastic lures.

Thermal Shock Stress caused to fish by sudden changes in water temperature.

Threatened At risk of becoming endangered.

Threatened Species at risk of becoming endangered in the near future.

Tide The rise and fall of sea levels caused by gravitational forces.

Tippets Tippet is a thin, translucent strand of plastic that is made out of synthetic polymers (i.e., large, chain-like molecules).

Torsion Twisting of a body part.

Translocations Moving species from one location to another for conservation purposes.

Trawlers Large fishing vessels that drag nets through the water to catch fish.

Trekking Long, adventurous journeys on foot.

Tributaries Streams or rivers that flow into a larger river or lake.

Trolling or Drifting Trolling involves the use of live bait or artificial lures that are drawn through the water behind a slow-moving boat, originally rowed but now generally motor-powered.

Trolling Dragging a bait or lure behind a moving boat.

Trolling Trolling is a method of fishing where one or more fishing lines, baited with lures or bait fish, are drawn through the water at a consistent, low speed.

Twitching It is a technique where you cast out a heavier jig, usually 3/8th to 1/2 ounce, and let it sink to the zone where the silvers are holding.

Two Hook Bottom Rig A fishing rig with two hooks for bottom fishing.

Unbarbed Lacking barbs, smooth.

Under Sized Fish Fish that are smaller than the legal size limit for catching.

Uplands High or hilly land.

Upstream The direction against the current of a stream.

Upstream The direction against the current of a stream.

Vomerine Teeth Teeth located on the vomer, a bone in the roof of the mouth of some fish.

Vulnerable Species that are at high risk of becoming endangered.

Vulnerable Susceptible to harm or damage.

Water Hyacinth An invasive aquatic plant that can clog waterways and affect local ecosystems.

Watershed An area of land that separates waters flowing to different rivers, basins, or seas.

Wave Height The vertical distance between the crest and trough of a wave.

Welded Joined or fused together using heat or pressure.

Wobbler Spoon A type of fishing lure that wobbles to mimic the movement of prey.

MIX
Papier aus verantwortungsvollen Quellen
Paper from responsible sources
FSC® C105338

If you have any concerns about our products,
you can contact us on
ProductSafety@springernature.com

In case Publisher is established outside the EU,
the EU authorized representative is:
Springer Nature Customer Service Center GmbH
Europaplatz 3, 69115 Heidelberg, Germany

Printed by Libri Plureos GmbH
in Hamburg, Germany